T0296264

CAMBRIDGE LIBRARY COLLECTION

Books of enduring scholarly value

Life Sciences

Until the nineteenth century, the various subjects now known as the life sciences were regarded either as arcane studies which had little impact on ordinary daily life, or as a genteel hobby for the leisured classes. The increasing academic rigour and systematisation brought to the study of botany, zoology and other disciplines, and their adoption in university curricula, are reflected in the books reissued in this series.

A Naturalist in Western China with Vasculum, Camera and Gun

Ernest Henry Wilson (1876–1930) was introduced to China in 1899 when, as a promising young botanist, he was sent there by horticulturalist Henry Veitch (1840–1924) to collect the seed of the handkerchief tree, *Davidia involucrata*, for propagation in Britain. Subsequent trips saw Wilson bringing back hundreds of seed samples and plant collections, introducing many Chinese plants to Europe and North America. He wrote extensively about his travels in China: this two-volume work was published in 1913. Although much of the text is concerned with plant life, Wilson also gives a great deal of attention to the wider landscape around him. In addition, Wilson took a camera, and these volumes contain photographs of parts of China rarely seen by Europeans in the early twentieth century. Volume 1 covers his travels from Hupeh (Hubei) to Szechuan and into the Tibetan region before ending at Wa Wu Shan.

Cambridge University Press has long been a pioneer in the reissuing of out-of-print titles from its own backlist, producing digital reprints of books that are still sought after by scholars and students but could not be reprinted economically using traditional technology. The Cambridge Library Collection extends this activity to a wider range of books which are still of importance to researchers and professionals, either for the source material they contain, or as landmarks in the history of their academic discipline.

Drawing from the world-renowned collections in the Cambridge University Library, and guided by the advice of experts in each subject area, Cambridge University Press is using state-of-the-art scanning machines in its own Printing House to capture the content of each book selected for inclusion. The files are processed to give a consistently clear, crisp image, and the books finished to the high quality standard for which the Press is recognised around the world. The latest print-on-demand technology ensures that the books will remain available indefinitely, and that orders for single or multiple copies can quickly be supplied.

The Cambridge Library Collection will bring back to life books of enduring scholarly value (including out-of-copyright works originally issued by other publishers) across a wide range of disciplines in the humanities and social sciences and in science and technology.

A Naturalist in Western China with Vasculum, Camera and Gun

Being Some Account of Eleven Years' Travel

VOLUME 1

ERNEST HENRY WILSON

CAMBRIDGE
UNIVERSITY PRESS

CAMBRIDGE UNIVERSITY PRESS

Cambridge, New York, Melbourne, Madrid, Cape Town,
Singapore, São Paolo, Delhi, Tokyo, Mexico City

Published in the United States of America by Cambridge University Press, New York

www.cambridge.org
Information on this title: www.cambridge.org/9781108030458

© in this compilation Cambridge University Press 2011

This edition first published 1913
This digitally printed version 2011

ISBN 978-1-108-03045-8 Paperback

A NATURALIST
IN WESTERN CHINA

THE AUTHOR

A NATURALIST IN WESTERN CHINA

WITH VASCULUM, CAMERA, AND GUN

BEING SOME ACCOUNT OF ELEVEN YEARS' TRAVEL,
EXPLORATION, AND OBSERVATION IN THE MORE
REMOTE PARTS OF THE FLOWERY KINGDOM

BY

ERNEST HENRY WILSON, V.M.H.

WITH AN INTRODUCTION BY

CHARLES SPRAGUE SARGENT, LL.D.

WITH ONE HUNDRED AND ONE
FULL-PAGE ILLUSTRATIONS AND A MAP

VOL. I

METHUEN & CO. LTD.
36 ESSEX STREET W.C.
LONDON

First Published in 1913

TO

MY WIFE

PREFACE

IN the following pages I have endeavoured to give a general account of Western China, more especially of its natural history and of the manners and customs of the non-Chinese peoples inhabiting the Chino-Thibetan borderland. The attempt is based as broadly as possible, and it is earnestly hoped that the information will be of interest to many sorts and conditions of people.

My travels in Western China began early in 1899, and had for their object the collecting of botanical specimens and the introducing of new plants into the gardens of Europe and North America. I have made four separate expeditions, covering in all nearly eleven years, and the nature of my work made it necessary for me to eschew the beaten tracks of the Flowery Kingdom.

The opportunity to travel and study the natural history of China I owe to the business enterprise of the house of Veitch, the famous nurserymen of Chelsea, to whom I was recommended by Sir William T. Thiselton-Dyer, then Director of Kew Gardens, at the instigation of Mr. W. Watson, the present Curator of that establishment. My first two expeditions were in the interest of Messrs. Veitch ; the last two in that of the Arnold Arboretum of Harvard University. The results of these four trips are well known in the horticultural and botanical circles of Europe and North America.

In my wanderings in China I have been singularly fortunate. The Chinese treated me always with kindly courtesy and respect. I was in interior China

b

during the Boxer outbreak and the Russo-Japanese War, and visited places shortly before or after anti-foreign riots, but never experienced any incivility meriting the name. I engaged and trained as collectors a number of Chinese peasants, who served me faithfully throughout my journeys, and we parted with genuine regrets. At the commencement of my travels in China, Mr. Augustine Henry, now Professor of Forestry at Dublin, imparted to me much sound advice which I did my utmost to follow. To this gentleman and to the devoted services of my Chinese collectors must be largely attributed the results of my work in China.

It is exceedingly pleasant to recall the kindly acts and hospitality of the many people I have been privileged to meet during my wanderings. Exigencies of space forbid the mention of names but do not affect my sincere appreciation. But for meeting them one's life would have been very much the poorer and lonelier. To my friend, W. J. Tutcher of Hong-Kong, this book in part owes its inception, and to another friend, J. Hutson Edgar, I am indebted for much information concerning the peculiar customs of the Thibetans and other non-Chinese races.

In the preparation of this work, I have received much encouragement from Professor Charles S. Sargent, who has also contributed an introduction of the greatest value. To my friend Herman Spooner I am indebted for invaluable criticisms of the manuscript. To Walter R. Zappey, my associate on the third expedition, I owe much for assistance in details concerning the colours and measurements of the game-birds and mammals.

Two or three of the chapters I first published in the *Gardeners' Chronicle* during 1905-6, and that on insect white-wax in the *Chemist and Druggist*, 1906, but these have been remodelled to suit present requirements, and amended and corrected in accordance with increased knowledge.

With six exceptions, the illustrations are from

photographs taken by myself with a whole plate
Sanderson camera for the Arnold Arboretum, and
permission to use them I owe to Professor Sargent.
The photographs were developed and printed by Mr.
E. J. Wallis of Kew, who obtained from the negatives
the best possible results. For the illustration of
Budorcas tibetanus, I am indebted to Mr. Samuel
Henshaw, Director of the Museum of Comparative
Zoölogy at Harvard College.

It is quite impossible to record the full extent of
one's obligations, since much information is un-
consciously absorbed through contact with many
people and extensive reading. I should, however,
be lacking in filial respect did I not record my sense
of indebtedness to the Alma Mater who gave me both
inspiration and opportunity—the Royal Gardens, Kew.

ERNEST H. WILSON

THE ARNOLD ARBORETUM
HARVARD UNIVERSITY
July 1913

CONTENTS

xi

LIST OF ILLUSTRATIONS

LIST OF ILLUSTRATIONS

INTRODUCTION

THE botanical explorations carried on in China in recent years make it possible to compare the forest flora of eastern continental Asia north of about lat. 22° 30′ with that of eastern North America north of the Rio Grande. In these explorations Mr. Wilson has played an important part, and more than any other traveller has shown us the remarkable richness of the flora of western central China and the distribution and value of many of the most important Chinese trees. A comparison of the flora of eastern continental Asia with that of eastern North America made at this time cannot be entirely conclusive, for although much has been done to make known the Chinese flora much is still left undone ; and there are still vast regions of the Celestial Empire into which no botanist has as yet penetrated, and these may be expected to yield new harvests of still unknown plants.

It is not surprising that the forest flora of China is richer in genera than that of eastern North America, for although the area of the two regions under consideration is not very dissimilar there is a great difference in their topography. In eastern North America only a few mountain peaks reach an altitude of 6000 ft., and these are wooded to the summit. In China mountain ranges are more numerous, with peaks which sometimes rise far above the upper limits of vegetation, and on some of these mountain ranges the timber line is at least twice as high as the highest land in eastern America. The connection of the great mass of mountains of south-western China with the Himalayas which must be considered their western prolongation, and the great tropical region which extends uninterrupted by any large body of water southward from south-western China, will account for the presence in the Chinese flora of many Himalayan and tropical forms which have no counterpart in eastern North America. On the other hand, the flora of eastern North America has drawn from the large and arid plateau of Mexico many genera of Cactaceæ, the Agaves, Yuccas, Dasylirion, and other genera which have no representatives in China. While the

larger mountain systems, the greater height of land, and its more prolific neighbours can account for a larger number of genera in eastern
Asia than in eastern North America, it is not possible to find an
explanation for the greater number of species there of widely
distributed genera like Acer, Picea, Prunus, Sorbus, and Berberis,
which are more numerous in China than in any other part of the
world, or for the absence from eastern Asia of larger numbers of
species in genera like Cratægus and Amelanchier.

In eastern continental Asia there is nothing to compare with
the great maritime pine belt which extends from southern Virginia
to eastern Texas, and is one of the remarkable features of the flora
of eastern North America ; and the great forests of *Pinus Strobus* L.,
which once extended from northern New England and eastern
Canada to northern Minnesota, are but poorly replaced in north-
eastern Asia by trees of *Pinus koraiensis* S. and Z., scattered over a
comparatively restricted area in eastern Siberia and Korea. The
Black Oaks, with their lustrous leaves and biennial fructification,
which are so abundant and conspicuous, except in the extreme north,
all over eastern North America, are wanting in eastern Asia ; while
the Bamboos, the most widely distributed and the most generally
useful of all the forest plants of China, are represented in North
America by two small and unimportant species of Arundinaria
confined to the swamps and river bottoms of the southern states.

As a rule, to which, of course, there are a few exceptions, the
trees of eastern North America are larger and more valuable than
related Chinese species ; but of Chinese shrubs it can be said generally that they produce more beautiful flowers than the shrubs of
eastern North America, although to this statement there are also
some exceptions. A more detailed examination of the principal
groups of forest plants in the two regions will show the similarities
and the differences of the forest flora of the two regions.

CYCADACEÆ.—Four species of Cycas are found in southern
China, and in Florida the family appears in two species of Zamia.

CONIFERÆ.—This family is represented in China by fourteen,
and in eastern North America by nine, genera. In eastern North
America there are only two genera which are not also represented
in China, Taxodium, which is replaced there by the nearly related
Glyptostrobus and Chamæcyparis, represented, however, in Japan
by two important trees. Libocedrus, Cupressus, Cunninghamia,
Pseudolarix, Keteleeria, and Fokienia have no eastern American
representative. In Pinus, eastern North America, with fifteen
species, has the advantage of eastern continental Asia, in which

INTRODUCTION

only eight species occur; and in eastern America Pine trees are individually larger, more numerous, and more generally distributed than in China. In Picea, China, with its twenty species, has a decided advantage over eastern America, where only three species occur, and in Abies, China, with its nine species, is richer than eastern North America, where there are only two species, and of these one is found only on the highest peaks of the southern Appalachian Mountains. Of Tsuga there are two species in eastern North America and two species in China, but *Tsuga canadensis* Carr. is a larger tree and much more widely distributed than any of the Chinese species. Larix, on the other hand, is better represented in eastern continental Asia, where it is widely distributed with several species from eastern Siberia to the mountains of Western China, where it sometimes forms large forests, while in eastern America there is a single species confined to the north-eastern part of the continent and is a small tree which southward is found only in swamps. Juniperus is represented in China by six species, and in North America by five species; but none of the Chinese Junipers are as large or as widely distributed as *Juniperus virginiana* L.; and none of them produce so valuable wood as that species and *J. barbadensis* L. Thuya is represented in each region with a single species of about equal importance. In eastern America Taxodium is a large and valuable timber tree widely distributed in the South Atlantic and Gulf regions, while its Asiatic representative, Glyptostrobus, is a small tree confined to the banks of a few streams in south-eastern tropical China.

TAXACEÆ.—In this family the advantage is all with China, with Taxus, Torreya, Cephalotaxus, Gingko, and Podocarpus, while in eastern America it appears with only a single species each of Taxus and Torreya, small trees found only in a few small isolated groves in western Florida.

GNETACEÆ. — Represented in continental Asia by Ephedra and Gnetum, this family does not appear in eastern North America.

PANDANACEÆ.—One species of Freycinetia and two species of Pandanus represent this Old World family in southern China.

PALMÆ.—About the same number of species of Palms are reported from the two regions, fourteen species in seven genera in China, and sixteen species in eight genera in eastern North America. The species in the two regions belong to different genera, with the exception of *Cocos nucifera* L., which is found on all the tropical shores. In eastern North America Palms extend farther north than in China, and some of the dwarf species cover in the southern United States great areas of dry sandy land with almost impenetrable thickets, as dwarf Bamboos make travel on some of the mountain slopes of Western China almost impossible.

LILIACEÆ.—Of this family Heterosmilax is found only in eastern Asia, but Smilax occurs in the two regions, in each of which it is about equally represented by a number of species ; Yucca and Dasylirion are American.

ARACEÆ.—Plants of this family are sometimes woody in southern China, where species of Pothos and Rhaphidophora are large climbers, but in eastern North America all the members of this family are herbaceous.

PIPERACEÆ.—Piper, with woody species in southern China, is the only genus of this family in the two regions.

CHLORANTHACEÆ.—Of this small tropical family Chloranthus, in China, is the only representative in the two regions.

SALICACEÆ.—In the number of species of Populus in the two floras there is no great difference. None of the Asiatic species grow to a larger size, perhaps, than the American Cottonwood (*Populus deltoidea* Marsh.), but with this exception the Asiatic species are larger and more valuable trees than the American species, notably the Manchurian *Populus Maximowiczii* A. Henry, and the north China *Populus tomentosa* Carr., which are among the largest and most beautiful Poplar trees of the world. In Salix there is probably no great difference in the number of species in the two regions, although there is still much to be learned of the alpine species of Western China. In eastern North America Willows are mostly shrubby, only three or four species attaining to the dignity of small trees, while in eastern Asia there are probably ten or twelve arborescent species, and some of these are trees of considerable size.

JUGLANDACEÆ.—In this family the advantage is with eastern continental Asia, with four genera against two in eastern North America, where there is no representative of Pterocarya, Engel-hardtia, or Platycarya. Juglans is common to the two regions, but Carya is not known in China ; and the presence in eastern North America of this genus in many widely distributed species, valued as timber trees and for the nuts produced by some of the species, economically, at least, makes up for the absence of the three genera which occur in China and not in eastern North America. Juglans, in eastern America with three species, and continental Asia, with four species, is of nearly equal importance in the two regions. None of the Asiatic species, however, compare in size with the American *Juglans nigra* L., but *Juglans regia* L., whose original home is now believed to be on the mountains of northern and Western China, yields the most valuable nuts and the most valuable timber produced by any species of the genus.

LEITNERIACEÆ, a family of a single species of Leitneria, is North American.

MYRICACEÆ.—Comptonia is confined to eastern North America, but Myrica, with a small number of species, occurs in the two regions, with one species, *Myrica Gale* L., common to them both. In North America there is no species which at all resembles *Myrica rubra* Lour. with its edible fruits.

BETULACEÆ.—In eastern North America Carpinus appears as a small widely distributed tree, but in continental Asia seven or eight species of the two sections, into which this genus has been separated (Eucarpinus and Distigocarpus), are common. Ostrya is represented in each region by a single species, the eastern American Ostrya being much more generally distributed and more abundant than the Chinese species, which appears to be confined to the mountain forests of Hupeh and Szechuan. The monotypic Ostryopsis is confined to Mongolia and China. Of Alnus there are five species in eastern North America, four of these being shrubs and one a small tree ; but in eastern continental Asia there are at least six or seven species of this genus, and one of these, *Alnus cremastogyne* Burk., is a large tree sometimes 100 ft. high, shading the banks of many streams in Western China with groves of splendid specimens. Betula forms a considerable part of the forests of eastern Siberia, and is common on many of the mountain ranges of China, especially those in western Szechuan, where it reaches altitudes of 10,000 ft. In eastern Asia, however, there is no species which, like *Betula nigra* L. of eastern North America, can thrive on the banks of streams in the nearly tropical heat of regions like Florida, Louisiana, and Texas, where this tree grows to its largest size. The number of species in the two regions is not very different, and as timber trees the Birches of one region are probably as valuable as those of the other. It is doubtful, however, if any eastern Asiatic Birch tree ever grows to the size sometimes attained by *Betula lutea* Michx., of the forests of north-eastern North America.

FAGACEÆ.—In eastern North America there is a single species of Fagus ranging from eastern Canada to Florida and Texas, and one of the largest and most common trees of all this region. In eastern continental Asia Fagus does not extend into the north, and appears to be confined to the mountain forests of the central western provinces, in which three species are now known ; these are smaller and less important trees than the American Beech.

Castanea is more important in the number of species in eastern continental Asia than in North America, but it is doubtful if any Asiatic species is anywhere as common or forms such a large part of the forest as the American *Castanea dentata* Borkh. forms on some Appalachian slopes ; and in height and girth of trunk this American tree has no Asiatic rival.   In eastern North America there are two other species ; of these one is a small tree or shrub and the other a shrub, both bearing a single nut in the involucre. In Western China there are also two species with similar fruit, but one of these, *Castanea Vilmoriniana* Dode, is a noble tree and the largest of the eastern Asiatic Chestnuts ; the other is a small shrub, to be compared with the American *Castanea nana* Muhl. The Japanese *Castanea crenata* S. and Z. reaches Korea, and *Castanea mollissima* Blume, another small tree, ranges from the neighbourhood of Peking to the mountains on the Thibetan border.   Castanopsis, which is related closely to Castanea, has its headquarters in south-eastern Asia, with several species in southern China, and one in California, but no representative in eastern North America.   It is possible that the number of species of Quercus is greater in eastern continental Asia than in eastern North America. Oak trees, however, are much less widely distributed in the former region and are not numerous at the north, the Chinese Oaks being chiefly confined to the southern provinces and usually evergreen. Some of these evergreen Oaks should be referred to Pasania, distinguished from the true Oaks by the arrangement of their flowers in bisexual aments, the pistillate in several-flowered clusters below the staminate, like the flowers of Castanea and Castanopsis.   As has been already stated, there are no Black Oaks in China, and no species which are counterparts of the eastern American Chestnut Oaks.   The northern White Oaks are inferior in size to several of the White Oaks of eastern North America, and it is doubtful if any of the southern evergreen species equal in size *Quercus virginiana* Mill., the Live Oak of the southern United States.

ULMACEÆ.—No Elm tree of eastern Asia equals the so-called American Elm (*Ulmus americana* L.) in size and beauty, but it is probable that the genus Ulmus has a larger representation of species in western continental Asia than it has in eastern North America, although it is still impossible to speak with much knowledge of Chinese Elm trees, which are very imperfectly understood. It is interesting that the section of the genus (Microptelea), which flowers in the autumn, has representatives in the two regions, two in eastern America and one in China, the only other species of this group growing on the Himalayas. The monotypic Planera occurs only in eastern North America, and Pteroceltis and Zelkowa have no American representatives. Celtis is common to the two regions, but the trees of this genus are larger and appear to be more generally distributed in eastern North America than in eastern continental Asia. The tropical genus Trema is represented in both regions.

MORACEÆ.—Of this family the monotypic Maclura is American, and Broussonetia, Cudrania, and Artocarpus, which occur in China, have no American representatives. There are two species of Morus in each of the two regions, but neither of the two American Mulberries compare in value with the Chinese *Morus alba* L., for the leaves of this tree and its numerous varieties furnish the best and chief food of the silkworm in all countries where silk is made. In Ficus the advantage is with China, both in the number of species and in the size of individuals, only two species having secured a foothold in tropical Florida, where they are comparatively small trees. Its nearness to tropical south-eastern Asia, which is one of the great centres of distribution of this genus, will account for the presence of some forty species of Ficus in southern China, where some of the species grow to a very large size.

PROTEACEÆ.—Helicia, in south-eastern China, is the only genus represented in the floras of the two regions.

LORANTHACEÆ.—In this family Phorodendron is North American; Arceuthobium is North American and Japanese, and Loranthus and Viscum are found in China and not in eastern North America.

SANTALACEÆ.—Pyrularia and the monotypic Darbya have not been found in China. Henslowia is eastern Asiatic, and Buckleya has a representative in the floras of the two regions.

OPILIACEÆ.—One species of the tropical genus Cansjera, in south-eastern China, is the only woody member of this family in our two regions.

OLACACEÆ.—Of this tropical family Schœpfia and Ximenia have reached southern Florida from the West Indies, each with a single species ; in China it appears only in Schœpfia.

ARISTOLOCHIACEÆ.—This family is represented in the floras of the two regions by the genus Aristolochia, with more numerous woody species in China than in eastern North America.

POLYGONACEÆ.—No arborescent or shrubby species of this family is reported in China, but in tropical Florida two species of Coccolobis occur, and a species of Brunnichia is widely scattered through the southern United States.

NYCTAGINACEÆ.—*Pisonia aculeata* L., an inhabitant of tropical shores in many parts of the world, probably reaches south-eastern China, as it occurs in Formosa ; with other species of this genus it is common in tropical Florida.

TROCHODENDRACEÆ.—This family is not represented in the flora of eastern North America but appears in China in Euptelea.

CERCIDIPHYLLACEÆ and EUCOMMIACEÆ, too, have no American representatives, but appear in Western China each with a monotypic genus, Cercidiphyllum and Eucommia.

RANUNCULACEÆ.—In this family Pæonia occurs in China but not in eastern North America, and the monotypic Xanthorrhiza is Appalachian. Clematis is common to both regions, with a much larger number of species in eastern Asia than in eastern North America, where the genus is poorly developed.

LARDIZABALACEÆ is an Asiatic and Chilian family, with Decaisnea, Stauntonia, Holbœllia, Akebia, Sinofranchetia, and Sargentodoxa in China.

BERBERIDACEÆ.—The woody plants of this family are much more numerous in China than in eastern North America. Mahonia and the monotypic Nandina do not occur in the latter region, where there is only one species of Berberis, while in eastern continental Asia, which must be considered the headquarters of the genus, some forty species are now recognized.

MENISPERMACEÆ.—The woody members of this family are represented in eastern North America by Menispermum and Cocculus. These occur also in western continental Asia, where Sinomenium, Diploclisia, Stephania, Cyclea, Tinospora, Limacia, and the monotypic Pericampylus are also interesting members of this family.

MAGNOLIACEÆ.—Magnolia is represented in the two floras by about the same number of species. In China, however, species occur in two groups, one of which produces its flowers before the appearance of the leaves, and in the other the leaves are nearly fully grown before the flowers open. To the latter group all of the American species belong. Some of the American Magnolias are larger trees than the Chinese species, and no Asiatic Magnolia compares in beauty with *Magnolia grandiflora* L. of the southern United States or equals *Magnolia macrophylla* Michx. in the size of leaves and flowers. Liriodendron appears in each region with a single species, but the American representative of this genus is a larger and much more widely distributed tree. Illicium and Schisandra appear in the two regions, the former with three species in China and one in the south-eastern United States, and the latter with one American and eight Chinese species. Michelia, Kadsura, and the monotypic Manglietia and Tetracentron are Chinese, the latter being one of the largest and most interesting of the Chinese trees.

CALYCANTHACEÆ.—Calycanthus, with several species, is eastern North American only, and Meratia (Chimonanthus), with two species, is Chinese and does not appear in eastern North America.

ANONACEÆ.—This tropical family reaches eastern North America with several species of Asimina, its most northern representative, and with Anona in tropical Florida. Uvaria, Artabotrys, Unona, Polyalthia, and Melodorum represent it in south-eastern tropical Asia.

LAURACEÆ.—In eastern North America this great, mostly tropical, family appears only in Persea, Ocotea, Sassafras, Litsea, Lindera, and Misanteca, but in eastern Asia there are eight genera of Lauraceæ, Cryptocarya, Beilschmiedia, Cinnamomum, Machilus, Sassafras, Litsea, Lindera, and Cassytha. Of Sassafras there is a species in each region, but the American species is a much more widely and generally distributed tree, the Chinese Sassafras being confined to the mountain slopes of western Hupeh and Szechuan. Litsea, which appears in eastern North America in one small shrub, in China is represented by at least a dozen species, among them several small trees. Lindera, too, is more important in eastern continental Asia than in eastern North America, where there are only two shrubby species, while China can boast of nearly ten times as many species ; some of these are large trees. The greater wealth of China in plants of this family appears, too, in the important genus Cinnamomum, including the species which yields the camphor of commerce, and in Machilus, of which several species are large and valuable timber trees.

CAPPARIDACEÆ.—This family appears in the tropics of the two regions with Capparis and Cratæva. Capparis is common to both, but Cratæva of south-eastern China has no American representative.

NEPENTHACEÆ.—One species of Nepenthes represents, in southern China, this Old-World family of a single genus.

SAXIFRAGACEÆ.—This family occurs in each of the two regions, with Philadelphus, Hydrangea, Decumaria, Itea, and Ribes. Deutzia (with one species in Mexico), Cardiandra, Schizophragma, Pileostegia, and Dichroa occur in China but not in eastern North America, which has no woody genus of the family not found also in eastern continental Asia.

PITTOSPORACEÆ.—Pittosporum, which reaches southern and Western China with a few species, is the only genus of this family in the two regions.

HAMAMELIDACEÆ.—This family is more important in the number of genera in eastern continental Asia than in eastern North America, where there is one endemic genus, Fothergilla. Hamamelis and Liquidambar occur in the two regions, and in China the family is represented also by Distylium, Corylopsis, Fortunearia, Sinowilsonia, Loropetalum, Sycopsis, Eustigma, Rhodoleia, and Altingia. In each region the Liquidambar is a large, widely distributed, and valuable timber tree. The Chinese Hamamelis, like one of the American species and the species from Japan, flowers in the winter.

PLATANACEÆ.—Platanus, the only genus of the family, which is represented in eastern North America by a large, common, and widely distributed tree, has not reached eastern Asia.

ROSACEÆ.—Of the thirty-four genera of the woody plants of this family found in the two regions, Neillia, Stephanandra, Sorbaria, Sibiræa, Exochorda, Cotoneaster, Osteomeles, Chænomeles, Docynia, Pyrus, Eriolobus, Pyracantha, Rhaphiolepis, Eriobotrya, Photinia, Stranvæsia, Rhodotypos, Kerria, Prinsepia, Pygæum, and Maddenia occur in China only. Three genera, Aronia, the monotypic Neviusia, and Chrysobalanus are American and not Chinese ; and ten genera are common in the two regions, Physocarpus, Spiræa, Rosa, Malus, Sorbus, Amelanchier, Cratægus, Rubus, Potentilla, and Prunus. Of the genera common to the two regions, Physocarpus, with one species in eastern Siberia, is better represented in eastern North America, where the genus is widely distributed with several species. On the other hand, the closely related Spiræa has a few small eastern American species, but abounds in China, which is the centre of greatest distribution of this genus. Eastern continental Asia, too, is greatly superior to eastern North America in species of Rosa, and in their variety and horticultural value, for China is the home of *Rosa lævigata* Michx., *Rosa bracteata* Wendl., *Rosa Banksiæ* R. Br., *Rosa multiflora* Thunb., *Rosa indica* L., the origin of the Tea Roses of gardens, and of *Rosa rugosa* Thunb. The number of species of Malus is probably about the same in the two regions, but it is interesting that those of eastern North America all belong to a group (Coronariæ) which is not represented in eastern Asia, where the small-fruited species with a deciduous calyx predominate. Sorbus in eastern North America is represented by two species of the Aucuparia section, while in eastern Asia there are nearly thirty species in this group and at least ten species of Aria which does not appear at all in the flora of eastern North America. Amelanchier, which is very widely distributed through eastern North America, with a number of species, of which two are small trees, has but one shrubby Chinese species. In Cratægus the difference between the floras of the two regions is even more remarkable. In all of eastern continental Asia only twelve species have been found; in eastern North America are more forms of Cratægus than of any other genus of plants, and probably a thousand species. In Rubus the difference in the number of species in the two regions is probably not great; several of the American species, however, produce more valuable fruit than any of the Asiatic species. *Potentilla fruticosa* L. appears in the two regions with two other related species in eastern Asia and one in eastern North America. The composition of Prunus is unlike in the two regions. Of the true Plums (Prunophora) there is only a single species in eastern continental Asia (*Prunus salicina* Lindl.), confined to southern and Western China, no plum tree being found anywhere in the north; in eastern North America plum trees have a wide distribution from the valley of the St. Lawrence River to Florida and Texas, a larger number of species occurring in the Arkansas-Texas region than in any other part of the world. Padus, on the contrary, is represented in eastern North America by only four species, while in eastern continental Asia about seventeen species are recognized. None of these, however, equal the American *P. serotina* Ehrh. in size or in value as a timber tree. Laurocerasus appears in eastern North America in two species and in eastern China in three species. Cerasus has but three eastern North American representatives and a much larger number in eastern continental Asia; and Amygdalus, Persica, and Armeniaca occur in eastern Asia and not in eastern North America.

CONNARACEÆ.—Rourea in southern China is the only representative of this family in the two regions.

LEGUMINOSÆ.—Of the genera of woody plants of this family the following occur in eastern North America but not in eastern continental Asia: Lysiloma, Prosopis, Parkinsonia, Cercidium, Amorpha, Eysenhardtia, Robinia, Coursetia, and Ichthyomethia; and the following in eastern continental Asia and not in eastern North America: Fordia, Ormosia, Millettia, Maackia, Caragana, Clitoria, Pueraria, Rhynchosia, Dalbergia, Euchresta, Mezoneurum, Cæsalpinia, Pterolobium, Entada, and Albizzia. The following genera have representatives in the two regions: Pithecolobium, Acacia, Leucæna (probably naturalized in southern China), Mimosa,

Cercis, Cassia, Gleditsia, Gymnocladus, Sophora, Cladrastis, Wisteria, Erythrina, Desmodium, Lespedeza, Dalbergia, and Sesbania.

RUTACEÆ.—In eastern Asia representatives of this family are certainly more important than those found in eastern North America, for they include Citrus, Limonia, Atalantia, two genera of interesting trees, Phellodendron and Evodia, besides Toddalia, Acronychia, Murraya, Clausena, Orixa, and Skimmia, while in eastern North America this family is represented only by Helietta, Ptelea, Amyris, and by Zanthoxylum which occurs also in the other region which contains the larger number of species.

ZYGOPHYLLACEÆ is represented in eastern North America by Guaicum and Porlieria, and in eastern continental Asia by Zygophyllum.

COCHLOSPERMACEÆ.—A species of Amoreuxia in southern Texas is the only member of this family in the two regions.

SIMARUBACEÆ appear in tropical Florida in a species of Simaruba and in Picrasma, and in Texas in Castela, while of this family China has given to the world one of its valuable trees in Ailanthus, and is also represented by Picrasma, Brucea, and Harrisonia.

BURSERACEÆ.—One species of Bursera is eastern North American, and the species of Canarium in China represent this family in the two regions.

MELIACEÆ.—Of the woody plants of this family in the two regions Swietenia is certainly the most valuable, although it is the only eastern American representative of the family ; while in China, representatives of this family are Aglaia, Amoora, Turræa, Cedrela, and Melia, one species of the last being widely and generally naturalized in the southern United States.

MALPIGHIACEÆ.—This family reaches tropical Florida with a species of Byrsonima and southern Texas with a species of Malpighia : its only genus in China is Hiptage.

POLYGALACEÆ.—Of this family only a few Chinese species of Polygala are frutescent, the species of this genus in eastern North America being herbaceous.

DICHAPETALACEÆ.—This small family is represented in the two regions by a single species of Dichapetalum in south-eastern China.

EUPHORBIACEÆ.—Woody plants of this family are more numerous in China than in eastern North America, where the following genera only appear : Andrachne, Drypetes, Croton, Ditaxis, Ricinella, Bernardia, Gymnanthes, Sebastiana, Stillingia, Hippomane, and Mozinna. These eastern North American representatives of the family are all small shrubs with the exception of Drypetes,

Gymnanthes, and Hippomane, which are small trees of tropical Florida. In eastern continental Asia woody plants of this family occur in Bridelia, Andrachne, Sauropus, Phyllanthus, Glochidion, Securinega, Breynia, Bischofia, Aporosa, Daphniphyllum, Antidesma, Microdesmis, Aleurites, Croton, Blachia, Claoxylum, Acalypha, Alchornea, Mallotus, Macaranga, Homonoia, Endospermum, Gelonium, Homolanthus, Erismanthus, Sapium, Sebastiana, and Excœcaria. Nearly all of these are tropical genera which, coming from the south, have obtained a foothold in south-eastern China. Only Andrachne, Croton, and Sebastiana have representatives in the two regions. Aleurites, a genus of trees which produce the wood-oil of commerce, is probably the most valuable genus of the family in the two regions.

BUXACEÆ.—One species of Pachysandra occurs in each of the two regions; the other genera of this family in the two regions, Buxus and Sarcococca, are Chinese.

CORIARIACEÆ, a family of a single genus, Coriaria, has a representative in China but not in America.

EMPETRACEÆ.—Empetrum occurs in north-eastern North America and in north-eastern continental Asia, and the other genera of this family, Corema and Ceratiola, are eastern North American, and have no Asiatic representatives.

ANACARDIACEÆ.—In this family, Pistacia, Rhus, and Cotinus are represented in the flora of the two regions. Metopium is American, and Spondias, Mangifera, and Dracontomelum are Chinese. In China, *Pistacia chinensis* Bunge is a large, widely distributed, and valuable tree, but in the United States *Pistacia mexicana* H.B.K. has secured only a precarious foothold on the northern bank of the Rio Grande in Texas. Of the members of this family in the two regions *Rhus verniciflua* Stokes, the Chinese Lacquer tree, is no doubt the most valuable.

CYRILLACEÆ.—This exclusively American family is represented in eastern North America by Cliftonia and Cyrilla.

AQUIFOLIACEÆ.—Of this family, Ilex is widely distributed in the two regions, and the monotypic Nemopanthus is east North American. Ilex usually grows to a larger size in China than in eastern North America, but is less northern in its range in the former region where most of the species are evergreen.

CELASTRACEÆ.—In this family, Celastrus and Evonymus are common to the two regions. Tripterygium, Perrottetia, Microtropis, and Elæodendron are eastern Asiatic and not American, and Pachystima, Maytenus, Crossopetalum, Gyminda, Schæfferia, and Mortonia are American. Only one species of Celastrus and three species of Evonymus occur in eastern North America, but in eastern continental Asia the species of these genera are much more numerous, and the species of Evonymus are usually larger and more beautiful plants.

HIPPOCRATEACEÆ.—Of this small tropical family there is a species of Hippocratea in tropical Florida and two or three in southern China.

STAPHYLEACEÆ.—Represented in China by Staphylea, Turpinia, Euscaphis, and Tapiscia, this family is much more important in eastern continental Asia than in eastern North America, where there is a single species only of Staphylea.

ICACINACEÆ.—Without an eastern North American genus this family appears in China in Iodes, Mappia, and the monotypic Hosiea.

ACERACEÆ.—Eastern continental Asia with its sixty-four species is far richer in Acer than eastern North America, where only ten species occur. The American Maples, however, are more widely distributed, and are larger and more valuable timber trees; Dipteronia and Dodonæa of this family are Chinese.

HIPPOCASTANACEÆ.—Of this family, Æsculus appears in two arborescent species in China, one in the north and one on the mountains of the west, but in eastern North America, where more species are segregated than in any other part of the world, four arborescent and four shrubby species occur in the southern United States. The monotypic Bretschneidera is Chinese.

SAPINDACEÆ.—Of the woody plants of this family found in our two regions, Urvillea, Serjania, Exothea, Hypelate, Cupania, and the monotypic Ungnadia are American, and the monotypic genera Xanthoceras and Delavaya, with Nephelium, Schmidelia, Kœlreuteria, and Pancovia, are Chinese; Sapindus is common to the two regions.

SABIACEÆ.—Without representatives in eastern North America, this family appears in China in Sabia and Meliosma.

RHAMNACEÆ.—Of this family, several genera reach tropical Florida from the West Indies and the dry region of Texas from Mexico, and the number is larger in eastern North America than in eastern continental Asia. The exclusively American genera are Rhamnidium, Reynosia, Ceanothus, Condalia, Karwinskia, Colubrina, and Gouania; and the Asiatic genera are Ventilago, Paliurus, and the monotypic Hovenia. Sageretia, Zizyphus, Berchemia, and Rhamnus have representatives in the two floras. Species of

Rhamnus, however, are more numerous in eastern continental Asia than in eastern North America, and *Rhamnus davuricus* Pall., and other Chinese species, from which a green dye is made, are more valuable than any of the American species.

VITACEÆ.—Of the Grape family, three genera, Tetrastigma, Cayratia, and Leea, occur in China and not in eastern America; and Ampelopsis, Parthenocissus, and Vitis are common to the two regions; Cissus reaches tropical Florida but has not been reported from southern China. Species of Vitis are less numerous in eastern North America than in eastern continental Asia; and in North America there is no species which corresponds with the spiny-stemmed Grape vines of China (Spinovitis).

ELÆOCARPACEÆ.—The forest flora of the two regions is only represented by the Asiatic Elæocarpus and Sloanea (Echinocarpus) of this family.

TILIACEÆ.—Tilia is widely distributed in the two regions with rather more species in the Asiatic region. In size and in value as ornamental trees there is not much difference between the American and Asiatic Lindens. The Asiatic genera Grewia, Corchoropsis, and Triumfetta do not appear in eastern North America.

MALVACEÆ.—In eastern North America there are woody species in Pavonia, Hibiscus, and Thespesia, and in western continental Asia only in Urena, Hibiscus, and Abutilon.

BOMBACEÆ.—Only the Asiatic Bombax represents this family in the woody plants of the two regions.

STERCULIACEÆ.—The large genus Sterculia has several Chinese representatives, including *Sterculia platanifolia* L. f., now naturalized in several of the southern United States. Of this family these genera also appear in China: Heritiera, Reevesia, Kleinhovia, Helicteres, Pterospermum, Abroma, and Buettneria, among which are several large trees, while in eastern North America are only Hermannia, Melochis, and Nephropetalum, all small shrubs of arid Texas.

DILLENIACEÆ.—Unrepresented in eastern North America, the family appears in China in Tetracera, Actinidia, and Clematoclethra.

THEACEÆ.—Much more important in eastern continental Asia than in eastern North America where only Gordonia and Stewartia occur, this family has several woody plants in China, including Thea, Gordonia, Stewartia, Schima, Ternstrœmia, Eurya, Hartia, Tutcheria, and Adinandra. One of the species of Thea, from the leaves of which tea is made, is the most important member of the family;

and in another (Camellia) are found some of the most valued and generally cultivated ornamental trees and shrubs.

GUTTIFERÆ.—Of this family only Ascyrum, Hypericum, and Clusia appear in eastern North America, but in eastern continental Asia, Ascyrum, Hypericum, Cratoxylon, Garcinia, and Calophyllum represent this family.

TAMARICACEÆ.—With Tamarix and Myricaria in eastern continental Asia this family has no representative in eastern North America, although one species of Tamarix is occasionally naturalized in the southern United States.

CISTACEÆ.—Hudsonia of the Atlantic coast region is the woody representative of this family in the two regions.

COCHLOSPERMACEÆ and KŒBERLINIANÆ.—A shrubby species of Amoreuxia of the former, and a species of Kœberlinia of the latter, both in Texas, are the only members of these families in the two regions.

CANELLACEÆ.—A West Indian species of Canella which has reached tropical Florida is the only member of this family in the flora of the two regions.

FLACOURTIACEÆ.—Without a representative in eastern North America this family contributes some of its most interesting trees to the Chinese flora in Xylosma, the monotypic Carrieria, Itoa, and Idesia, and in Poliothyrsis.

STACHYURACEÆ, of which Stachyurus is the only genus, is Asiatic.

TURNERACEÆ.—A shrubby species of Turnera from southern Texas is the representative of this family in the two regions.

PASSIFLORÆ.—Although Passiflora appears in several herbaceous species in the southern states, *Passiflora ligulifolia* Mast. of southern China is the only woody species of the family in our two regions.

CARICACEÆ.—*Carica Papaya* L., now naturalized in many of the warm countries of the world, is possibly a native of southern Florida.

DAPHNACEÆ.—In this family the advantage is with eastern continental Asia, as Dirca is its only American representative, while in China there are species of Daphne, Edgeworthia, Wickstrœmia, and Aquilaria.

ELÆAGNACEÆ.—In this family Shepherdia is American, Hippophaë Chinese, and Elæagnus is found in the two regions.

LYTHRACEÆ.—The monotypic Decodon is the only woody plant of this family in eastern North America, while in China appear species of Lagerstrœmia and Woodfordia, and the monotypic Lawsonia.

RHIZOPHORACEÆ.—Rhizophora is common on the shores of tropical Florida, but the family is more largely represented in tropical China by Kandalia, Bruguiera, and Carallia.

COMBRETACEÆ.—Represented in tropical Florida by Bucida, Conocarpus, and Laguncularia, this family appears in China in Combretum, Quisqualis, and Illigera.

MYRTACEÆ.—Rhodomyrtus, Eugenia, Psidium, and Bæckea of the Myrtle family have reached south-eastern China, while in tropical Florida occur Eugenia, Anamomis, Chytraculia, and Syzygium.

MELASTOMACEÆ.—Of this family woody species of Barthea, Allomorphia, Blastus, Bredia, Anplectrum, and Memecylon occur in China, but a species of Tetrazygia is the only woody member of the family in eastern North America.

ARALIACEÆ.—An arborescent species of Aralia and a species of Echinopanax are the only tree and shrub of this family in eastern North America; in eastern continental Asia the family is more largely represented by Aralia, Acanthopanax, Fatsia, Nothopanax, Heptapleurum, Dendropanax, Heteropanax, and Hedera.

CORNACEÆ.—Of this family Garrya occurs in eastern North America but not in China, where are found the monotypic Camptotheca, a large tree, Davidia, Alangium, Helwingia, Torricellia, Marlea, and Aucuba. Nyssa and Cornus are common to the two regions. Nyssa in America is widely and generally distributed from New England to Florida and Texas, with several species, of which two are large trees, but in China only a small tree is now known, confined to the central provinces. On the other hand, Cornus is more numerous in species in China than in eastern North America, six of the species at least being arborescent and one a tree occasionally 100 ft. high.

CLETHRACEÆ.—Of this family there are three species of Clethra in each of our two regions.

ERICACEÆ.—In the number of genera of this family eastern North America has the advantage of the Asiatic region, with twenty-three genera in the former and only seventeen in the latter. Such genera as Bejaria, Leiophyllum, Menziesia, Kalmia, Zenobia, the monotypic Oxydendrum, Gaylussacia, and Arctostaphylos have no eastern Asiatic representatives. Enkianthus, Craibiodendron, and Diplycosia are Chinese, without American representatives, and the following genera are common to the two regions : Vaccinium, Gaultheria, Chamædaphne, Loiseleuria, Phyllodoce, Andromeda, Arctous, and Rhododendron. No single genus except Cratægus so well illustrates, perhaps, the differences in the floras of the two regions as Rhododendron. In eastern North America there are only six species of true Rhododendron, all confined to the extreme eastern part of the continent, and, with one exception, of restricted range, but some one hundred and sixty species have already been distinguished in eastern continental Asia, where the genus is widely distributed and where, on the mountains of the western and south-western provinces, is the greatest segregation of these plants in the world. On the other hand, only three species of Azalea have been found in China, while in eastern North America, which is the region of their greatest development, ten or twelve species are recognized.

THEOPHRASTACEÆ.—The only member of this family in the floras of the two regions is a single species of Jacquinia in tropical Florida.

MYRSINACEÆ.—This family has a much larger representation in eastern continental Asia than in eastern North America, only one species each of Ardisia and Rapanea having reached tropical Florida, while in southern China there are several species of Ardisia, Rapanea, and Myrsine, and where also Mæsa, Embelia, and Ægiceras occur.

PLUMBAGINACEÆ.—With Plumbago in the two regions the family is also represented in China by Ceratostigma.

SAPOTACEÆ.—In this family eastern North America has the advantage with six genera, while only three genera reach southern China. Of these only Sarcosperma is not represented in the American flora. The other genera which are found in China, Sideroxylon and Chrysophyllum, occur in tropical Florida, which has been reached also by Dipholis and Mimusops, while Bumelia, which is an American genus, is widely distributed through the southern United States with several species.

EBENACEÆ.—Of this family only one genus, Diospyros, is represented with two species in eastern North America, and eight or nine species in China. As a fruit tree one of the Chinese species is much more valuable than the North American species.

STYRACEÆ.—Of this family Styrax occurs in the two regions. Halesia, with three species, is eastern North American ; and the monotypic Alniphyllum and Pterostyrax are Chinese.

SYMPLOCACEÆ.—Symplocos, the only genus of the family, appears with one species in the southern United States, and is largely represented in China, where twenty species are distinguished.

OLEACEÆ.—In this family also the advantage is with eastern continental Asia, with eight genera, while only four are eastern North American. Fraxinus, Chionanthus, and Osmanthus are common to the two regions, Adelia is American only, and Fontanesia, Forsythia, Syringa, Ligustrum, and Jasminum are Chinese and not American. Fraxinus is widely distributed in each of the two regions, with probably about the same number of species in each, but the American species are usually larger and more valuable timber trees. As an ornamental plant the American Chionanthus is superior to the Chinese representative of the genus, but China's contributions to gardens from this family in Forsythia, Syringa, Ligustrum, and Jasminum more than make up for the beauty of the American Chionanthus.

LOGANACEÆ.—Gelsemium, with one species, and Buddleia, with a species of southern Texas, are the only woody representatives of this family in eastern North America. These genera occur in China with Strychnos, Gærtnera, and Gardneria.

APOCYNACEÆ.—Vallesia, Thevetia, and Trachelospermum are woody plants of this family in eastern North America. It has a larger representation in southern China in Plumeria, Melodinus, Rauwolfia, Alyxia, Alstonia, Parsonsia, Pottsia, Wrightia, Ecdysanthera, Anodendron, Trachelospermum, and Scindechites.

ASCLEPIADACEÆ.—Roulinia, with a species of southern Texas, is the only woody plant of this family in eastern North America, but in eastern continental Asia occur woody species of Pentaneura, Cryptolepis, Periploca, Taxocarpus, Calotropis, Holostemma, Graphistemma, Metaplexis, Henrya, Gymnema, Marsdenia, Stephanotis, Pergularia, Dregea, and Hoya.

CONVOLVULACEÆ.—Ipomœa and Argyreia are Chinese representatives of this family, and a woody species of Ipomœa has reached Florida from the tropics.

BORRAGINACEÆ.—Cordia, Bourreria, and Ehretia are the North American genera of this family, with woody species. Cordia and Ehretia appear in China in a larger number of species than in eastern North America, and Tournefortia is Asiatic and not American.

VERBENACEÆ.—Aloysia, Lantana, Citharexylon, Duranta, Callicarpa, and Avicennia, represent this family in the southern United States. In China, Callicarpa, with several species, Premna, Gmelina, Vitex, Clerodendron, Caryopteris, Sphenodesma, and Avicennia make a larger representation of the family, only Callicarpa, with one American species, and Avicennia occurring in the two regions.

LABIATÆ.—A few small shrubs of Salvia in Texas, and the genus Elsholtzia, with several species in eastern continental Asia, represent the woody plants of this family in our two regions.

SOLANACEÆ.—Species of Lycium and Solanum, which occur in each of the two regions, are the only woody plants of this family.

SCROPHULARIACEÆ.—Of this family Leucophyllum, a small shrub of western Texas, is the only woody plant in eastern North America, but in eastern continental Asia it is represented by the important genus Paulownia, of several species of large trees, and by Brandisia.

BIGNONIACEÆ.—In this family Campsis and Catalpa are common to the two regions. The monotypic Chilopsis, and Aniso-stichus and Crescentia occur only in the southern United States. Oroxylum, Dolichandrone, Stereospermum, and Radermachera are Chinese and not American.

GESNERACEÆ.—Æschynanthus and Lysinotus, with woody species in China, are the representatives of this family in the two regions.

MYOPORACEÆ.—A species of Myoporum of southern China is the representative of this small family in the two regions.

RUBIACEÆ.—Of this family the American woody representatives are Cephalanthus, the monotypic Pinckneya, Exostemma, Genipa, Randia, Catesbæa, Hamelia, Guettarda, Erithalis, Chiococca, Strumpfia, Psychotria, Morinda, and Ernodea. Of these Cephal-anthus, Randia, Guettarda, Psychotria, and Morinda occur also in China, where also are woody representatives of Adina, Luculia, Wendlandia, Hedyotis, Mussænda, Adenosacme, Myrioneuron, Webera, Gardenia, Diplospora, Antirrhœa, Conthium, Ixora, Damnacanthus, Lasianthus, Pæderia, Hamiltonia, Leptodermis, Serissa, Emmenopterys, Dunnia, Pavetta, and Uncaria.

CAPRIFOLIACEÆ.—Of the ten genera of woody plants of this family found in the two regions, Dipelta, Leycesteria, Kolkwitzia, and Abelia (with a species in Mexico) occur only in China. Of the other genera, Sambucus, Viburnum, Symphoricarpos, Linnæa, Lonicera, and Diervilla are common to the two regions. Symphori-carpos is chiefly American, with a single Chinese species, and Vibur-num, with some seventy species, is richer in China than in eastern continental America, although the American species grow to a larger size and are more ornamental. Lonicera is poorly represented in eastern North America with twelve species, while in eastern continental Asia more than one hundred species are recognized, the region of their greatest segregation being on the mountains of the central and western provinces.

GOODENIACEÆ.—A Chinese species of Scævola is the only woody plant of this family in the two regions.

COMPOSITÆ.—Iva and Baccharis, of the Atlantic coast region, are the American shrubby representatives of this family, which occurs in China in several species of Blumea and in Pertya.

It appears, therefore, that 129 natural families are represented in the two regions under consideration ; that of these 92 are common to the two regions, that 12 occur in eastern North America, but not in eastern continental Asia, and that 25 occur in eastern continental Asia and not in eastern North America. Of the 692 genera in the two regions 155 are common to both, while 158 are found in eastern North America and not in eastern continental Asia, and 379 are found in eastern continental Asia and not in eastern North America. Of the tropical genera 76 have reached southern Florida and 89 south-eastern China.

From Mexico the flora of the United States has derived 42 genera of the woody plants of Texas.

It is impossible to form an accurate estimate of the comparative number of species of woody plants in the two regions at this time, but including Cratægus it is probable that the number is as great in eastern North America as in eastern continental Asia.

C. S. SARGENT

ARNOLD ARBORETUM
July 1913

A NATURALIST IN WESTERN CHINA

CHAPTER I

WESTERN CHINA

Mountain Ranges and River Systems

WESTERN China is separated from Thibet proper by a series of parallel mountain ranges running almost due north and south, and divided by narrow valleys. On some maps the name Yun-ling is applied to the whole system, with sections marked Hsueh shan, Hung shan, Taliang shan, and so on. A great many local names, the majority of them unpronounceable when converted into English, are also applied to this system, but outside certain maps no one general name for it exists. Later we shall have much to say about this region, for the time being it suffices to note the general trend of the ranges and a few of their important features.

Made up largely of razor-backed ridges, following one another in quick succession, these ranges are separated by narrow valleys, or rather ravines. The higher peaks are well above the snow-line, and for height, savage grandeur, and wondrous scenery are comparable only to the Himalayan alps of India. The whole region is practically uncharted and unsurveyed, and it is the author's firm conviction that some of the peaks rival in altitude the greatest of the Himalayan giants.

About lat. 33° N., in the neighbourhood of Sungpan Ting, a mighty spur is thrown out from these ranges of perpetual snow, and extends, with a slight southerly dip, due east for some 10° of longitude, terminating in low hills near Anluh

Hsien, in north-eastern Hupeh. This spur appears on maps under the general name of Kiu-tiao shan (nine mountain ridges), Ta-pa ling, or Ta-pa shan. The two latter names have direct reference only to important peaks of the spur, and the first is the most appropriate, since it denotes a series of parallel chains closely packed together. The Kiu-tiao shan forms the boundary between Szechuan and the northern provinces of Kansu and Shensi, and is the watershed between the middle Yangtsze and the Han River systems. Attaining its greatest altitude in the neighbourhood of Chengkou Ting, long. 108° 30', lat. 32° 15' N. (approx.), it radiates from this climax buttress-like spurs in all directions. Those on the south form the boundary between Szechuan and Hupeh and extend downwards beyond the Yangtsze River. Subsidiary spurs and others thrust out from more easterly points of the range, make the whole of north-western Hupeh exceedingly wild and rugged. In the middle of the province the Yangtsze River has forced itself through these spurs, which run at right angles to its course, and formed the famous Yangtsze Gorges.

Another spur, or rather series of spurs, not so clearly defined as the preceding, and of less altitude, is thrown out in the neighbourhood of Tali Fu, long. 100° E., lat. 25° 42' N. (approx.), in western Yunnan. It extends across northern Yunnan, southern Kweichou, and northern Kwangsi, and forms the boundary between Hunan and Kiangsi on the north and Kwangtung on the south. In eastern Kiangsi it is deflected north and north-north-east, finally reaching the sea in the neighbourhood of Ningpo, long. 121° 35' E., lat. 29° 50' N. (approx.). This mountain system extends across some 21 parallels of longitude, and forms the watershed between the Yangtsze River on the north and numberless rivers on the south. Of these the Red River, reaching the sea in the Gulf of Tongking, and the West River, which enters the sea near Macao and Hongkong, are the chief.

Innumerable lateral spurs are given off by this system, and the country is extremely broken, especially in the western parts, with which we are concerned. The province of Kweichou is one mass of mountains, and the same is true of southern Hupeh and southern Szechuan. In these three

A PEAK OF THE BARRIER RANGE (TA-P'AO SHAN), CIRCA 21,000 FT.

areas there are subsidiary ranges of considerable altitude dipping in various directions and connected up with spurs to form a heterogeneous and complex mountain system. The outstanding feature of the whole region west of 112th parallel of longitude is the entire absence of plain or plateau, or anything in the nature of flat, level country, with the solitary exception of the area forming the Chengtu Plain. Of this we shall speak in due course. East of the 112th parallel the Yangtsze River flows through a flat, alluvial plain in which isolated, or more or less connected, mountain ranges and spurs occur, but with this region we are not in this work concerned.

The most important region comprised within the mountain systems above described and west of the 112th parallel is that termed by Richthofen the " Red Basin of Szechuan." This region includes the whole of Szechuan east of the Min (Fu) River to near the Hupeh boundary. It is a region of vast agricultural wealth, with a magnificent river system, teeming with large cities, towns, and villages, and supporting an enormous population. With the solitary exception of cotton, which is imported from the coast, it is self-contained, with a surplus of produce to spare for export. Salt is produced in unlimited quantities in very many districts ; coal, iron, and other minerals of economic importance abound. In short, the " Red Basin " is one of the richest and fairest regions in the Chinese Empire.

The whole of Western China, with which this work is concerned, lies within the Yangtsze River basin. According to the geographical information at present available, the Yangtsze has its source almost due north of Calcutta, in latitude about 35° N., on the south-east edge of the Central Asian steppes. Its exact length is unknown, but it is estimated to exceed 3000 miles. From its source it pursues a tortuous course, nearly due south, through wild and partially unknown country for 1000 miles. Then suddenly turning eastward it flows right through the heart of China for some 2000 miles, finally reaching the sea immediately to the north of Shanghai.

From its mouth to Ichang, 1000 miles, it is navigable for steamers at all seasons of the year, though in winter difficulties in the way of shoals and sand-bars are encountered. The greatest difficulty is experienced between Hankow and Ichang, and this section is operated by a small fleet of shallow draught steamers specially built for the trade. The regular steamer-fleet plying between Shanghai and Hankow is also specially designed for the service and is luxuriously fitted. Ocean-going steamers of deep draught can ascend as far up as Hankow, except in low-water season. In summer the river overflows and invades much of the low-lying country contiguous to its course, and the chief difficulty in navigation at such times is to keep to the channel. The difference between summer and winter level is very considerable and varies to a large extent, according to the width of the river and the nature of its banks. At Ichang the river is 1100 yards from bank to bank, and the average difference between summer and winter levels is about 40 feet ; in the gorges which commence some 5 miles west of Ichang, the river is narrowed to a third of its usual breadth and the difference exceeds 100 feet. Above Ichang the river is obstructed by rapids, rocks, and other impedimenta, and is navigated by specially built native boats that range up to 80 tons displacement. The difficulties of navigation are more especially confined to the stretch of the river between Ichang and Wan Hsien, a distance of about 200 miles. From Wan Hsien to Pingshan Hsien, some 500 miles farther west, the navigation becomes easier.

Much has been written on the possibility of opening the Yangtsze River to merchant steamer traffic from Ichang westwards. So long ago as April 1900, two British river gunboats of shallow draught, small in beam and length, and of a special design, ascended as far as Chungking, the commercial capital of Western China, distant above Ichang some 400 miles. Later these boats ascended as far west as Pingshan Hsien and one of them succeeded in reaching Mei Chou, a city on the Min (Fu) River, about 140 miles above its junction with the Yangtsze at Sui Fu. Since this exploit two larger and more powerful British gunboats have been built for this work and are now stationed at Chungking, which has been made a naval base. France and Germany, following the British lead, have also gunboats stationed at Chungking. During suitable seasons these craft move up and down the river, and regularly every year one or more visit Pingshan Hsien and Kiating Fu, the latter city being about 100 miles north of Sui Fu on the Min River.

The advent of the gunboats had been anticipated early in 1898 by a small launch called the *Leechuan*, commanded by Captain C. Plant and owned by the late Mr. Archibald Little, the pioneer foreign merchant of these regions. The experimental test made by this launch took practical shape in 1900 when a commercial steamer named the *Pioneer*, captained by Plant and operated by a British syndicate, in which Mr. Little figured, was placed on this service. She made a trip prior to the Boxer outbreak, after which she was chartered by the British Government and was finally purchased for naval purposes.

On 27th December 1900, a German merchant ship, the *Suihsiang*, specially designed and built for the purpose, left Ichang for Chungking, but was wrecked and totally lost below the Tungling Rapid only some 40 miles above Ichang.

Early in 1910 the task was again taken in hand; this time a powerfully constructed tug named the *Shutung*, towing alongside a flat for passengers and cargo, was employed. This outfit, owned by a Chinese syndicate, was commanded by the same Captain C. Plant. The venture proved successful, and fourteen round trips were made during the year. It is fitting that the man who pioneered the whole business should succeed in demonstrating the practicability of merchant steam navigation on the Upper Yangtsze. The work, however, is dangerous, exceedingly difficult, and, moreover, costly, and unless some improvements are made in the river-bed, it will be some time before any considerable fleet of steamers ply on these waters.

Above Pingshan Hsien navigation is only practicable for small native craft in certain short interrupted sections. The river flows for the most part through gorges or between steep mountains, and its course is frequently broken by dangerous rapids and cataracts that produce a seething, foaming swirl in which nothing can live. In the autumn of 1911 an adventurous French naval officer made an extraordinary journey down the Upper Yangtsze to Sui Fu in native boats specially built for the purpose. An account of this journey should prove exciting reading.

It has been mentioned that the most difficult stretch of the Middle Yangtsze was that between Ichang and Wan Hsien. This is the region of the world-famous Yangtsze Gorges. Five in number, these gorges extend from the immediate west of Ichang to Kuichou Fu, a distance of about 150 miles. Throughout this stretch the river flows between perpendicular walls of rock, is narrowed to a third or less of its usual width, and becomes in consequence very deep. Soundings taken by the British gunboats in their ascent in 1900 gave 63½ fathoms of water in two places, and this when the water at Ichang was rather less than 6 feet above zero mark ! The cliffs, composed largely of hard limestone, are 500 to 2000 feet or more high, and commonly 500 to 1000 feet or more sheer. The scenery hereabouts is savagely grand and awe-inspiring.

Foreign maps without exception give the name of Yangtsze kiang (variously spelt) to this magnificent river. So far I have never met a Chinese to whom this name is intelligible. I have read that the name denotes " Son of the Ocean," and is applied to the section between Wuhu and the sea. This may be so, I have no knowledge on the point. Many local names are given to stretches of this river, but from Sui Fu, in western Szechuan, to its mouth it is universally spoken of by Chinese as the Ta kiang (Great River), occasionally it is rendered Chang kiang (Long River), or simply Kiang, meaning *The* River. West of Sui Fu it is called the Kinsha Ho (River of Golden Sands); the Chinese do not consider this the main stream, but regard the Min (Fu) River as the principal. They recognize that the Kinsha has the larger volume, but it is navigable only for some 40 miles and then loses itself in wild and barbarous regions. The Min, on the other hand, is navigable to Chengtu, some 200 miles above Sui Fu, and is therefore to the utilitarian mind of the Chinese of much greater importance. From near Batang northward the Kinsha Ho is known by the Thibetan name of Drechu, and finally near its source it goes under the Tangut name of Murussu.

In ascending the Yangtsze from Ichang to Chungking the observant traveller is struck by the insignificant character of the tributary streams. Apparently the only one of importance joins the main stream at Fu Chou on the right bank. This

THE YANGTSZE RIVER AT ICHANG

stream, the Kien kiang, rises in western Kweichou and flows through the heart of this province. It is navigable from its mouth to Szenan Fu, and beyond, for specially constructed native boats. Apart from this river there is no tributary of seeming importance until Chungking is reached, yet nearly every town and village of note stands at the junction of some small stream with the Yangtsze. Here and there men will be found hauling small, stout-bottomed boats over the stones at the mouths of these small rivers. That the main stream is joined by many tributary streams a glance at the map proves. In western Hupeh the country is wildly mountainous, and the streams are torrents, pure and simple. In eastern Szechuan the country is much less wild and the streams of a different character, and why they appear unnavigable is, on the surface, not obvious.

In 1910 I journeyed overland by a little-known route from Ichang to Chengtu. Entering Szechuan a little to the north of Taning Hsien, I travelled due west to Paoning Fu, and from thence south-west by the main road. On this journey I crossed all the principal streams which join the Yangtsze on its left bank east of the Min River. The surprising thing observed was the fact that they were one and all navigable for boats of varying sizes for long distances. On inquiry, I found that navigation ceased on most of them some 2 to 5 miles before their union with the Yangtsze. The Kuichou Fu, Yunyang, and Kai Hsien Rivers may be cited as examples, affording evidence of this state of things.

Near the embouchure of tributary streams the Yangtsze is generally narrowed and the water gorged by boulders and detritus choking the mouths of these lateral waterways; rapids and races frequently occur at these points. The accepted view is that enormous quantities of debris are brought down by these tributaries and deposited at their mouths. This theory is all very well when applied to mountain torrents, but most of the streams under discussion pursue a comparatively placid course with easy currents for some 50 miles or more before reaching the Yangtsze. Their volume and force of current is insufficient even in summer floods to carry down the enormous quantities of detritus which choke up their mouths. My

8 A NATURALIST IN WESTERN CHINA

personal observations put the responsibility on the main stream itself. During the summer floods the Yangtsze brings down vast quantities of mud and detritus, which it deposits wherever opportunity offers. Flowing as the Yangtsze does, more or less between steep banks, the mouths of tributary streams afford the most favourable places for the deposition of this debris. The volume of the main stream is enormously greater, and its current so much stronger, than that of the tributaries that it simply thrusts them back and silts up their mouths. The small quantity of debris brought down by the tributary streams would also be deposited hereabouts owing to the slacking of the flow consequent upon the damming of their debouchure.

At Chungking the Yangtsze is joined on its left bank by the Kialing or so-called " Little River." A glance at a map shows that this river is made up of three streams which unite near Ho Chou. The Kialing River and its tributaries drain a fan-shaped area, in extent more than half of the entire Red Basin situated north of the Yangtsze. Their importance is due to their being navigable for such extreme distances. The most easterly branch is navigable, for small craft, to some 40 miles north of Tunghsiang Hsien ; the next branch is navigable to Tungchiang Hsien ; and the next to north of Pa Chou. The central (Paoning) river is navigable for fairly large boats to Kuangyuan Hsien, and skiffs laden with medicines and other native products descend to this town from Pikou, in the province of Kansu. The most westerly branch is navigable to Pai-shih-pu, a few miles north of Chungpa, and one of its western tributaries taps the north-east corner of the Chengtu Plain.

The Kialing River system is thus the greatest collecting and distributing waterway in Szechuan, and its commercial importance, probably greater than that of the Yangtsze itself and its tributaries west of Chungking, is not generally understood. The To kiang, which joins the Yangtsze at Lu Chou, though a natural stream, owes very much of its volume and importance to water artificially lead from Kuan Hsien across the northern part of the Chengtu Plain via Sintu Hsien, and a secondary branch via Han Chou, which meet together at the great market town of Chao-chia-to. In summer it is possible to descend in boats from Han Chou and Sintu Hsien to Lu Chou.

A GENERAL VIEW IN NORTH-WESTERN HUPEH

The Min (Fu) River proper, save at lowest water, is navigable from Kuan Hsien and Chengtu downwards. The Chengtu branch is artificially formed by canals led across the plain from Kuan Hsien, and unites with the Kuan Hsien stream and its tributaries at Chiangkou. A tributary of the Min, which joins at Hsinhsin Hsien, is navigable in high water for small boats to Kiung Chou, a city situated at the extreme south-west corner of the Chengtu Plain.

The Min (Fu) River rises some 35 miles north of the Sungpan Ting, near the boundary of north-west Szechuan and the Amdo region in lat. 33° N. (approx.). Immediately to the south of Sungpan city it plunges into wild, mountainous country, flowing through a gorge from which it emerges only a few miles north of Kuan Hsien, where it becomes navigable for rafts only.

At Kuan Hsien a famous and gigantic irrigation system is in operation, but of this we shall deal in due place.

The Min is really only a tributary of the Tung River, which it joins at Kiating Fu, but since it admits of navigation it is of more practical importance, and for this cause the Chinese give it pre-eminence. The Tung River is only navigable for a few miles above Kiating, though rafts descend from a much higher point west. Its tributary the Ya, which joins it immediately west of Kiating, is of greater commercial importance, and a very considerable raft traffic ascends and descends this stream from Yachou, which is the centre of the brick-tea industry of western Szechuan.

The Tung River is really one of the longest rivers in Szechuan, having its source in the north-eastern corner of Thibet, about lat. 33° 40′ N. It flows through the western frontier of the tribes country, where it is known as the Tachin Ho (Great Gold River), and ultimately strikes the highway from Chengtu to Lhassa at Wassu-kou, a hamlet 18 miles east of Tachienlu. From this point to its union with the Min at Kiating it is called the Tung Ho, though around Fulin it goes by the name of Tatu Ho. Owing to its unnavigability its commercial importance is small, but this does not excuse the geographer's scant appreciation of it in the past, even if it explains the Chinese view.

Considerably west of Pingshan Hsien the Yangtsze is

joined by the Yalung River, an artery almost equal in volume to the main stream itself. This river rises in the north-eastern limits of the Thibetan highlands in the same general country as the Yangtsze, but to the south-east. It flows more or less due south throughout the whole course, but the region it traverses is, if anything, less known than that through which the Yangtsze flows. In its upper parts it is called the Niachu, since it flows through the country of the Niarung tribes, and its map cognomen, "Yalung," is probably a transliteration of the name Niarung.

On its right (south) bank the Yangtsze receives many streams rising in northern Yunnan and Kweichou, but none equalling in importance those uniting on the left bank. However, all are significant factors in the distribution of merchandise in these parts, even though geographically they are of comparatively small moment. The important thing to be remembered in connection with the river-system here mentioned is this—the Yangtsze River is the main artery of China in general and Western China in particular, but Szechuan owes its agricultural wealth and general prosperity principally to the Kialing and Min Rivers with their network of navigable contributary streams and canals. The rivers west of Sui Fu flow through wild, mountainous, sparsely populated regions, and their course is so much obstructed that practically no boats ply on their waters and even ferries are scarce.

CHAPTER II

WESTERN HUPEH

General Topography and Geology

THE country comprising western Hupeh, with which we are concerned, lies west of the 112th parallel of longitude. The city of Ichang, situated on the Yangtsze River, just west of this parallel and about 1000 geographical miles from the mouth of the river, is a convenient starting-point for exploring this region. This important town is a treaty port, opened to foreign trade in 1877. The population is roughly estimated at 30,000. There is also a small foreign community consisting of a British Consul, Imperial Maritime Customs' staff, a few business men, and missionaries of Roman Catholic and various Protestant denominations. There is very little local trade, but Ichang being practically the head of steam navigation on the river, is a most important transhipping port. Six steamers regularly trade between Ichang and Hankow, and the thousands of native craft lined up in tiers attest its importance as an entrepôt of trade. In the near future it is destined to be a most important junction on the Hankow-Chengtu railway and already important work on this enterprise has been commenced there. Ichang is well known, and every year foreigners visit it in increasing numbers intent on seeing the famous gorges which lie immediately beyond. It is easily reached from Shanghai and from Peking. From Shanghai palatially fitted and specially designed steamers leave every night for Hankow, 600 miles up the Yangtsze. From Peking and Hankow a good express train runs to and fro weekly. Between Hankow and Ichang a regular fleet of steamers keep up constant communication. Ascending the river by steamer from Hankow the hilly country commences about 40 miles below Ichang. At first low, the hills gradually increase in height, and by the time Ichang is reached one is fairly among the mountains. In the vicinity of the town the hills are pyramidal in outline, with prominent cliffs near by ; north, south, and west of the town the country is much cut up, forming an archipelago of peaks 2000 to 4000 feet high, the peaks themselves being offsets from spurs attaining altitudes of 7000 to 9000 feet, situated some days' distance beyond. These pyramidal hills around Ichang are very interesting and never fail to attract the attention of travellers. They are made up of a substratum of pebbly conglomerate, on which are reared thin, horizontally deposed strata of marine limestone, red shale and sandstone, over-capped with sandy clays. The strata are piled with great regularity, and when erosion is equal on all sides the characteristic pyramidal shape is produced and maintained. This formation is general from the edge of the great plain to Ichang, and occasionally it contains thin beds of coal. It is of comparatively recent age, dating back to Permo-Mesozoic times. The dominant fossils it contains are Cycads, and the youngest rocks probably belong to the Oölitic series. The cliffs and bold peaks to the north, south, and west of Ichang are made up principally of Paleozoic limestones, with a little shale and sandstone, the latter of the Mesozoic period. The strata are folded in apparent conformity and are without notable metamorphism. In eastern Szechuan these rocks extend beneath the Red Basin. The Yangtsze has forced itself right through them and formed a series of mighty chasms in which the structure of the various formations is beautifully exhibited.

In the neighbourhood of Hwangling Miao (30 miles west of Ichang), and westwards for 10 miles to the Tungling Rapid, granitic gneiss is exposed. These are the oldest rocks in this region and the only Pre-Cambrian formation known *in situ* in the Middle Yangtsze. This section of the river is called the Ta-shih Ho (River of Dregs and Boulders), and well does it deserve this appellation.

The next oldest rocks of importance are those forming the cliffs opposite Nanto, in the Niukan Gorge and in the eastern half of the Wushan Gorge. This is a massive formation 4000 to

A VIEW NEAR HSINGSHAN HSIEN WITH TERRACED FIELDS

5000 feet thick, in the major part composed of dark grey or liver-coloured limestone free from chert and containing both Cambrian and Ordovician fossils. It is, in fact, a great marine limestone in all its phases. It weathers into wonderful escarpments, often sheer for 1000 to 2000 feet, with slightly projecting summits, and is frequently many miles in extent. The cliffs on the right bank of the river opposite Nanto, which extend nearly to Hwangling Miao, are typical examples. At one of the major rapids during the low-water season, known as the Hsintan, some 45 miles west of Ichang, a bed of shale is beautifully exposed. This bed is some 1800 feet thick, and composed principally of olive-green argillite, with local black shale and quartzite. It is of the Middle Paleozoic age.

Resting apparently conformably on this series of shale is a vast deposit of Upper Carboniferous limestone 4000 or more feet thick. This is the characteristic formation throughout the Ichang and Mitan Gorges ; it occurs also throughout the western end of the Wushan Gorge and in the Kui Fu (Wind-box) Gorge beyond. The prevailing rock is dark grey or blackish limestone, full of marine fossils and with occasional thin layers of anthracite coal. This also weathers into wonderful escarpments, but commonly they are boldly rounded with less linear dimensions. This formation is the most general throughout western Hupeh, on both sides of the river, though greater on the north than on the south, where the Cambrian-Ordovician formation preponderates. Next in succession come the Permo-Mesozoic beds of red shale and sandstone, with thin layers of marine limestone and coal, which were described in reference to Ichang. These beds are characteristic of the country west of the Mitan Gorge, as far as the entrance to the Wushan Gorge, principally on the left bank. Coal occurs in this stretch in many places, more especially around Patung Hsien. Glacial deposits and signs of glacial action are in evidence in many parts of western Hupeh, though nowhere on a large scale. The most accessible of these is on the Yangtsze itself, opposite Nanto, a hamlet situated on the extreme western end of the Ichang Gorge and some 20 miles above Ichang city. At this point can be seen a glacial deposit about 120 feet thick, overlaid by marine limestone of the Cambrian-Ordovician age referred to above. All the evidences of ice action are well disclosed, and the whole deposit is most instructively exhibited. Since the deposition of these various systems great regional disturbances have taken place and the strata has commonly been bent up from a great depth. The summits of the very highest peaks in western Hupeh are usually comprised of Silurian (? Devonian) shales.

None of the useful or precious minerals occurs in quantity in western Hupeh. Coal is scattered through the entire region, but is nowhere found in abundance and the quality is indifferent. Iron ore is worked in places and in one or two localities the quality is good, but usually it is poor. Copper occurs in two districts (Chienshih and Hsingshan) but is not worked to any great extent. Salt, so abundant throughout the Red Basin of Szechuan, does not occur. The sandy clays and marls are used in brick and tile making, and lime is burnt in several places and used for building purposes. Both the clays and the limestone here mentioned belong to the Permo-Mesozoic beds. The carboniferous limestones are quarried and used for various construction works.

In the Gorges the main stream is joined by numerous lateral branches which flow through glens of wondrous beauty. These streams, winding their way through, usually fill nearly the entire bed of the glen and are bounded by walls of cliff 300 to 1000 feet sheer. Waterfalls are numerous and wherever it is possible vegetation is rampant. The tops of the cliffs are worn into curious and grotesque shapes. Caves abound and in these stalactites and stalagmites occur. Subterranean springs are common and many of the small rivers originate from such sources. They issue forth from some cave, or from the face of a cliff, or well up through level rocks. The Hsingshan River is an example of this mode of origin. The Chinese attach much legendary lore to all these caves and subterranean springs, and frequently associate fine temples with such spots.

In the vicinity of the Yangtsze the more commanding peaks and crags are crowned by temples, usually belonging to the Taouist cult. Commonly these temples cap seemingly inaccessible points, and one marvels how the material used in erecting them was transported thither. Whenever possible a few trees, usually *Xylosma racemosum*, var. *pubescens* (Winter-

SAN-YU-TUNG GLEN; CLIFFS OF CARBONIFEROUS LIMESTONE

green), Gleditsia, Cypress, Ginkgo, and Pine are planted near the temples and add much to the beauty of the scene. Such temples are well built, but unfortunately, since the interior is usually dark, filthy, and uninviting, a close inspection robs them of most of their charm. From the distance they look most picturesque, the style of architecture being in harmony with the surroundings, and one admires very much the taste and culture which called them into existence. The preservation of the " Good Luck " of towns, villages, and communities by the warding off of evil influences is a matter of great moment in China, and with this good work the temples are associated. The pagodas, found all over China, have been erected solely with this end in view. Geomancy enters very largely into Taouism and holds a most important place in Chinese thought, and, in fact, governs many of their actions. As illustrating this we will take an example at Ichang. Facing the town, on the right bank of the river opposite, is a pyramidal hill nearly 600 feet high, called by foreigners the " Pyramid." This hill was supposed to exert a baneful influence over the town, and was held responsible for the town's poverty in local *literati*. Not until a temple was built on an eminence behind the town, sufficiently high to enable it to overlook the Pyramid, was this evil influence counteracted, and the Goddess of Good Luck induced to smile on the town. The very year this temple was completed a student passed the provincial examinations with high honours. Was not this the beneficial result of the building ? The temple, called Tungshantzu, is richly endowed and forms a strikingly conspicuous object from all points of approach. The logic of " Fung Shui," as this cult is called, is beyond the grasp of the average Occidental brain, but of its effect on the Chinese mind one is constantly made familiar.

Too wild and savage for extensive agricultural development, and with a marked absence of useful mineral deposits, western Hupeh is one of the poorest, most sparsely populated, and least known parts of China. For these same reasons it is of particular interest to the botanist, since the vegetation there has been less molested than is usually found to be the case in China generally. Even here, it is hardly necessary to say, every available bit of land either is, or has been, under cultivation, but much of the country is of such a nature as to preclude agricultural development, even under Chinese patience and ingenuity.

Up to 3000 or 4000 feet, wherever it is possible, the mountain-slopes, hill-tops, and valleys are cultivated, but the country is made up so much of sheer cliff and crag, and is generally of such a rocky character, that even where cultivation is possible the crop won from the soil is poor and scarcely recompenses for the outlay of labour involved in its production. Above 6000 feet the higher slopes and mountains defy even Chinese skill and patience, and it is here that patches of virgin forest and much woodland remain. The higher mountains are rich in various Chinese medicines, and men eke out a livelihood in gathering them. Considerable areas in this higher country were formerly cleared and crops of the Irish potato raised. But some twenty odd years ago the potato disease attacked and devastated the crops, and ruined the peasants, who were forced to migrate to lower and more congenial altitudes. Ruined houses and numerous graves, overgrown with coarse herbs, brambles, and shrubs, tell of former habitation ; but to-day, in the higher parts of this region, it is possible to walk from morning till night without seeing an inhabited dwelling or a living person. Wherever the valleys admit of sufficient cultivation to support them, small riverine villages occur. Tiny hamlets, farmhouses, and peasants' huts are frequent up to 4500 feet altitude. Above this little agriculture is attempted and the population is exceedingly sparse.

I have travelled pretty extensively in the back-blocks of China during my eleven years' acquaintance with the country, and consider north-western Hupeh the most difficult part of China to explore, the Chino-Thibetan borderland not excepted. The absence of food-supplies and accommodation for coolies, the lack of roads, and the difficult nature of the country in general, render travel in this region exceedingly arduous.

As an Appendix to this chapter, it may be of interest to give an account of the flora obtaining in the vicinity of Ichang, which, from the amount of collecting that has been done in its neighbourhood, holds a classic place in the annals of botanical exploration work in China.

PRIMULA SINENSIS

APPENDIX

THE FLORA OF ICHANG

THE Flora of Ichang and the neighbourhood up to 2000 feet altitude, as included in this note, is essentially of a warm temperate character, and includes not a few sub-tropical forms. Nevertheless, we find also a number of cool temperate plants, and what really obtains is a fusion of these three floras, with the warm temperate element in the ascendancy. The following characteristic plants will serve to illustrate the point : *Aleurites Fordii, Liquidambar formosana, Ligustrum lucidum, Cæsalpinia sepiaria, Toddalia asiatica, Wisteria sinensis, Rhododendron indicum, Pyracantha crenulata, Primula sinensis, Anemone japonica, Aspidistra punctata, Reinwardtia trigyna,* and *Woodwardia radicans.* The low hills around Ichang are very barren-looking, being mostly clad with " spear grass " (*Heteropogon contortus*), with a few shrubs and herbs here and there, and relieved by small woods of *Pinus Massoniana* and *Cupressus funebris,* with occasional groves of the common Bamboo, *Phyllostachys pubescens.*

However, it is not to these low hills that we look for the floral wealth of Ichang, but to the limestone cliffs of the glens and gorges. Here the variety is astonishing, a striking feature being the quantity of well-known flowering shrubs.

The two first shrubs to flower in the early spring are *Daphne genkwa* and *Coriaria nepalensis.* It is a thousand pities we cannot succeed with the Daphne in England, since it is such a lovely plant—by far the finest species of the genus. Here, at Ichang, it grows everywhere, on the bare exposed hills, amongst conglomerate rock and limestone boulders, on graves, and amongst the stones which are piled around the tiny cultivated plats in the gorge, sometimes in partial shade, but more usually fully exposed to the scorching sun. The plants are, on the average, about 2 feet in height, and are but seldom branched. Imagine the annual suckers from a Plum tree, and you have the appearance of these Daphne plants.

18 A NATURALIST IN WESTERN CHINA

For two-thirds of their height they are so densely clad with flowers that they look like one large thyrse. The colour is lilac, often very dark ; but a white form is not uncommon. Its outward resemblance to Lilac leads to its being so called by the foreign residents at Ichang.

The Coriaria is not so well known and is not nearly so attractive. Its flowers are polygamous, and the plant when in fruit is rather showy. The Chinese consider its foliage and stem poisonous to cattle.

Wisteria sinensis is abundant, often scaling high trees, but the semi-bush form is the more common. Its flowers are borne in great abundance, and vary much in shade of colour, the white form being, however, rather rare.

Another well-known shrub which abounds here is *Loropetalum chinense*. On the tops of the cliffs, amongst loose conglomerate and limestone boulders, it forms a well-nigh impenetrable scrub. The bushes are seldom more than 3 feet in height, very much branched, and when in full flower look like patches of snow at a distance. Messrs. Veitch show the plant very well, but there is an enormous gulf between the best grown pot plants and the plants in a state of nature. In Devon and Cornwall, if planted in a rockery, it ought to thrive.

Rose bushes abound everywhere, and in April perhaps afford the greatest show of any one kind of flower. *Rosa lævigata* and *R. microcarpa* are more common in fully exposed places. *Rosa multiflora, R. moschata,* and *R. Banksiæ* are particularly abundant on the cliffs and crags of the glens and gorges, though by no means confined thereto. The Musk and Banksian Roses often scale tall trees, and a tree thus festooned with their branches laden with flowers is a sight to be remembered. To walk through a glen in the early morning or after a slight shower, when the air is laden with the soft delicious perfume from myriads of Rose flowers, is truly a walk through an earthly paradise.

In March and April *Sophora viciifolia* is very fine in the glens and gorges when it is covered with masses of bluish white flowers. This plant has a very wide distribution. It is common in Yunnan, and in the warm valleys of rivers bordering Thibet. The Ichang plant is much less spiny than that of Yunnan and western Szechuan. Possibly the latter is really the Indian *S. Moorcroftianum*. Two very common plants on the cliffs in the glens are *Eriobotrya japonica* (Loquat) and *Meratia præcox*. Both flower about Christmas. These are two out of many plants which formerly were erroneously supposed to be natives of Japan. Among conglomerate boulders *Caryopteris incana* is common, but is not nearly so fine as it is farther west. *Pyracantha crenulata* and *Vitex Negundo* are exceedingly common, and so also is *Cæsalpinia sepiaria*. This thorny shrub is semiscandent in habit, and very like the better known *C. japonica*. Its handsome foliage and erect thyrsoid racemes of bright yellow flowers make it a very conspicuous object.

Symplocos cratægoides, with its pretty white flowers and bright blue fruits, is abundant. This is a useful and charming shrub, and deserves to be better known. *Deutzia Schneideriana*, *Lagerstrœmia indica, Rhododendron indicum, Jasminum floridum*, *Nandina domestica, Ilex cornuta, Viburnum utile*, and *Buddleia officinalis* are all extremely common shrubs. Of other well or lesser known shrubs which are common, I may mention— *Abelia chinensis, A. parvifolia, Rhus Cotinus, Buddleia asiatica, Ilex pedunculata, I. corallina, Deutzia discolor, Desmodium floribundum, Elæagnus pungens, E. glabra, Spiræa chinensis, Eurya japonica, Hypericum chinense, Hydrangea strigosa, Berchemia lineata, Evonymus alata, Polygala Mariesii, Viburnum brachybotryum, V. propinquum, Thea cuspidata, Rubus parvifolius*, and many other species. *Chaenomeles sinensis* with red, and *C. cathayensis* with white or blush-white flowers, are commonly cultivated. Lengthy as is the list, I am not justified in omitting *Itea ilicifolia*. This Holly-like shrub, with long, pendent racemes of white flowers, is one of the handsomest of all the Ichang shrubs. Of fluviatile shrubs, the commonest are *Distylium chinense, Salix variegata, Ficus impressa, Rhamnus davuricus, Adina globiflora, Myricaria germanica*, and a curious Box (*Buxus stenophylla*). Climbers are very much in evidence, and include such beautiful plants as *Lonicera japonica, Trachelospermum jasminoides, Pueraria Thunbergiana, Clematis Henryi, C. Benthamiana, C. Armandi, C. uncinata, Vitis flexuosa, Parthenocissus Henryana, P. Thomsonii*, and *Mucuna sempervirens*.

This last is a rather remarkable plant. Two miles above Ichang, on the right bank, is an enormous specimen, called by foreigners the " Big Creeper." It covers several hundred square feet of ground, climbing over several Pine trees and many Bamboos. The base of the main trunk is almost as thick as a man's body ; the flowers are dark chocolate coloured, and are borne in racemes on the old wood ; the legumes are 2 to 2½ feet in length, and contain many large black bean-like seeds. It flowers in May.

Ichang does not possess a great number of trees, but the variety is really astonishing. *Paulownia Duclouxii* and *Melia Azedarach*, with their enormous panicles of flowers, are very striking in the spring. In the autumn, *Sapium sebiferum*, with its wonderful autumnal tints, stands alone. In winter the evergreen *Ligustrum lucidum*, and *Xylosma racemosum*, var. *pubescens*, are very conspicuous. The latter nearly always shelters some wayside shrine. Perhaps the commonest trees are—

Gleditsia sinensis, Rhus javanica, Platycarya strobilacea, Quercus serrata, Cedrela sinensis, and *Pterocarya stenoptera.* The Mistletoe occurs on the last-named tree. Other less common trees are *Sterculia platanifolia, Populus Silvestrii, Cratægus hupehensis, Celtis sinensis, Dalbergia hupeana, Acer oblongum, Cunninghamia lanceolata, Ailanthus glandulosa, Broussonetia papyrifera, Ulmus parvifolia, Hovenia dulcis, Sapindus mukorossi, Salix babylonica,* and *Sophora japonica.* Of this latter a curious variety occurs in which the leaves and young shoots are clothed with a dense white velvety indumentum.

As with flowering shrubs, so with herbs, though in a less degree, Ichang is the home of many favourite garden plants. One of the commonest and best known is *Primula obconica.* This charming herb abounds everywhere, but more especially in moist, grassy places on the banks of the Yangtsze and in the glens. Occasionally, under very favourable conditions, in height, size of flower, and luxuriance of foliage, it approaches the cultivated form, but more usually it is a dwarf and almost insignificant weed.

Again, Ichang is the home of the Chinese Primrose, and the type of the cultivated Chrysanthemum occurs there also. Other favourites which are common are—

Corydalis thalictrifolia, Anemone japonica, Sedum sarmentosum, Saxifraga sarmentosa, Iris japonica, Reinwardtia trigyna, Lycoris aurea, L. radiata, *Rehmannia angulata, Hemerocallis fulva,* and *H. flava.* Other characteristic herbs are : *Adenophora polymorpha, Bletia hyacinthina, Asarum maximum, Ophiorrhiza cantonensis, Viola Patrinii, Delphinium chinense, Lysimachia Henryi, L. clethroides, Potentilla chinensis, P. discolor, Fragaria indica, Thalictrum minus, Mazus pulchellus, Verbena officinalis, Platycodon grandiflorum,* and many *Compositæ, Leguminosæ,* and *Umbelliferæ.*

Perhaps Ichang is best known to horticulturalists generally as the home of the lovely *Lilium Henryi.* This acknowledged favourite occurs on the limestone and conglomerate rocks, but is now by no means common. *Lilium Brownii* and its varieties, *chloraster* and *leucanthum,* are fairly common ; *L. concolor* occurs, but is rare.

Ferns are not rich in species, but *Woodwardia radicans, Osmunda regalis, Pteris longifolia, P. serrulata, Nephrodium molle, Cheilanthes patula,* and *Gleichenia linearis* are very abundant. A variety of *Adiantum Capillus-Veneris* is very common on stalagmitic limestone in the glen. Pieces of these rocks covered with Ferns are detached and find their way all over China, being popularly known as " Ichang Fern-stones."

A hasty reference to the common floating plants of the ponds and ditches around Ichang must bring this note to a close. *Euryale ferox,* with its handsome foliage, is very common ; *Nelumbium speciosum* is, of course, cultivated. Other common aquatics are—*Limnanthemum nymphoides, Jussiæa repens, Salvinia natans, Trapa natans, Azolla filiculoides, Marsilea quadrifolia, Monochoria vaginalis, Eriocaulon Buergerianum,* and several species of Potamogeton and Utricularia. In late autumn, when the Azolla changes to a rich crimson tint, the ponds look very fine. In some rice fields near Ichang Dr. Henry found a very anomalous plant. It was made the type of a new genus—*Trapella sinensis,* and doubtfully referred to the natural order *Pedalineæ.*

CHAPTER III

METHODS OF TRAVEL

Roads and Accommodation

THE advent of steam navigation on the upper-middle Yangtsze has brought Chungking, the commercial metropolis of Western China, three weeks nearer the coast and occidental civilization. This is a very considerable gain to the would-be traveller in these regions, yet it only postpones for a little time longer the inevitable. Sooner or later the traveller must dispense with the comforts and luxuries of modern occidental methods of travel and adapt himself to those more primitive and decidedly less comfortable of the Oriental. In the regions with which we deal there is nothing in the nature of wheeled vehicular traffic save only the rude wheel-barrows in use on the Chengtu Plain. There are no mule caravans, and scarcely a riding pony is to be found. For overland travel there is the native sedan-chair and one's own legs ; for river-travel the native boat. Patience, tact, and abundance of time are necessary, and the would-be traveller lacking any of these essentials should seek lands where less primitive methods obtain. Endowed with the virtues mentioned, and having unlimited time at his disposal, he may travel anywhere and everywhere in China in safety, with considerable pleasure and abundant profit in knowledge. With her industrious toiling millions, her old, old civilization, her enormous natural wealth and wondrous scenery, China alternately charms and fascinates, irritates and plunges into despair, all who sojourn long within her borders. No country, outside Europe and North America, is of such perennial interest to the world at large as China. Ever-changing yet ever the same, she is the link which connects the twentieth century with the dawn of

OUR CHINESE HOUSE-BOAT

OUR CARAVAN

civilization, epochs before the Christian era. To travel leisurely through this vast country is an education which leaves an indelible impress on all fortunate enough to have had the experience. The Chinese do not see time from the Westerner's view-point, and for the traveller in the interior parts of China the first, last, and most important thing of all is to ever bear this in mind.

The majority of travellers still ascend the river above Ichang in native boats, and it will probably be a long time before a regular fleet of steamers ply these dangerous waters and render the native boat obsolete. The journey from Ichang to Chungking and beyond has been described so often that the subject is threadbare, and I have no intention of describing it over again. Volumes have been written on this subject, and some day perhaps a writer will arise and do full justice to the theme.

I have made the journey up and down many times, and on each occasion have been more and more impressed with the sublime beauty of the Gorges. The scenery in these savage chasms is all and more than any writer has described it as being. It must be seen to be fully understood and appreciated. The more often one travels up and down this stretch of the river the deeper grows one's awe and respect for the many rapids, swift currents, and innumerable difficulties which impede navigation.

The native boats are perfectly fitted for the navigation of these difficult waters ; they are the outcome of generations of experience, and the balance-rudder and turret-build have been used in these craft long before their adoption by Western nations. The men, too, who earn their livelihood in navigating these boats, understand their business thoroughly. Much has been written by hasty travellers on the shortcomings and incompetence of these men, that is as unwarrantable as it is undeserved. These Chinese boatmen are careful, absolutely competent and thorough masters of their craft, and the more one sees of them and their work the more one's admiration grows. Oriental methods are not occidental methods, but they succeed just the same ! When on the boat the Westerner will do well to adapt himself to Eastern methods ; any attempt to enforce those of the West generally ends in disaster. Many accidents on the Yangtsze have been caused through the foreigner, ignorant of local conditions, difficulties and dangers, forcing the captain of the boat to proceed against his better judgment. The traveller is advised when engaging a boat to do so through a responsible Chinese business house, to have an agreement drawn up setting forth the arrangements desired, and then to leave the boat-master to carry out his engagement in his own way. This is the only way to ensure safety, and on paper no one would attempt to gainsay it, yet in practice this is commonly done, but always to the jeopardy of the transgressor.

Since we shall have much to say on the subject of overland travel a word or two anent roads seems fitting and desirable. To the uninitiated this subject may seem trivial, but to the experienced it is otherwise. Chinese roads make a lasting impression on all who travel over them, and the vocabulary of the average traveller is not rich enough to thoroughly relieve the mind in this matter. The roads are of two kinds, paved and unpaved. I have yet to meet the traveller whose mind is thoroughly made up as to which of these is worse and the more difficult to negotiate. A clever writer once wrote : " An Imperial highway in China is not one which is kept in order *by* the Emperor, but rather one which may have to be put in order *for* the Emperor." [1] When any important official takes up duties in a distant part of the empire the local officials put the roads over which he has to travel in some semblance of repair. Such work is always hastily done by labour forced and grudgingly given, and in mountainous districts the first severe rainstorm destroys considerable portions of it.

It is nobody's real business to look after the roads, and nobody does. The land devoted to roadways is commandeered, and in agricultural districts the farmer takes good care to keep these roads down to a minimum width. It usually happens that the roadways get narrower and narrower every year, until the advent of some important official forces the local authorities into having them repaired and restored to their original width. Roads in China owe their origin to

[1] Arthur H. Smith, *Village Life in China*, p. 35.

the same causes that obtain elsewhere in the world, namely, military conquest and commercial interchange between distant localities.

Throughout the length and breadth of China run imperial highways, few in number, it is true, but of vast importance, since they connect the imperial capital with the capitals of the provinces. These were made for military purposes in early times, when the Emperors were busy conquering the country and extending their territories. They are all of great strategical importance, and were originally paved throughout with huge blocks of stone. Often, indeed, they were actually blasted and excavated from solid rock. They vary in width according to the configuration of the country and the nature of the traffic they have to carry. In the northern parts overland travel is commonly done by cart, and the roads are adapted to such traffic. In the parts with which we are concerned the country is too wild and rugged for wheels, and the only recognized mode of travelling is by means of sedan-chair. The imperial roads were originally made sufficiently wide to enable two chairs to pass one another freely. Ten to twelve feet is a broad highway in these parts, and it must be conceded that roads of such width amply serve their purpose. Unfortunately this width is rarely maintained for any considerable distance. The grading of these ancient highways was well done, and the whole work speaks volumes for the ability and energy of those old-time engineers. Like much else in China these roads were once magnificent, but to-day they are far from this. In general they are sadly neglected. Floods have destroyed them here and there, often the paving blocks have been stolen for house-building and other purposes, and gaps of unpaved, muddy stretches, almost impassable in rainy weather, occur all too frequently. Sufficient of the original road remains to stir admiration for the skill and foresight of the engineers, long since dead, and to set the traveller longing for those halcyon days of old.

In the prosperous parts of China, highways connect all the principal cities, town, and villages. These are usually 8 to 10 feet wide, and though originally paved throughout, are now in a state of more or less disrepair. Nearly all the towns and villages in Western China are situated on the banks of streams for the simple reason that the valleys offered lines of least resistance. Even when the streams are not navigable they afford easier means of access to the interior than the mountains and forest-clad country. In a general way all the older roads in China follow the courses of streams as closely as possible, leaving them only when the nature of the country necessitates the departure, and watersheds intervene.

Bypaths and narrow tracks permeate the country in every direction, and abound even in the most sparsely populated mountainous regions. Some one has very wisely made out that the exchange of salt was the first commerce engaged in by mankind at large. Salt is, and long has been, a Government monopoly in China, consequently the practice of salt-smuggling has gone on from time immemorial, and the majority of the mountain-paths were very probably first struck out by smugglers of salt. Indeed, many important trade-routes to-day, in China, presumably originated in this way. The province of Szechuan is abundantly rich in salt and also in mountain-paths. From a lengthy study I have come to regard this network of bypaths as the result of salt traffic, and more especially illicit traffic. There are to-day many such paths throughout the Hupeh-Szechuan boundary, used for practically no other traffic than that of salt, and by these paths salt still reaches certain districts in defiance of the law. Very useful, if difficult, the traveller finds these bypaths, for without them it would be impossible to traverse some of the wildest and most interesting parts of central and Western China.

When travelling overland in China it is not possible to use tents, and one has perforce to make use of such accommodation as the country affords. The Chinese do not understand tents, and it is unwise to try innovations in a land where the people are unduly inquisitive. The traveller gets along best when he avoids publicity as much as possible. On all the main roads there are inns of sorts, usually very filthy, and in season abounding in mosquitoes, creeping things, and stinks, the latter, in fact, being always in evidence. On the byways, and more especially in the mountains, accommodation is hard to find and is of the meanest description. However, one is usually tired,

HOSTEL AT CHE-TSZE-KOU, PINUS ARMANDI BEHIND

and any shelter suffices for a night's halt. In wet weather, or when held up through flooded torrents or what not, the absence of proper accommodation is acutely felt. In the wilds of China one hungers for the dâk bungalows of India and Ceylon, or accommodation on similar lines.

A traveller in China should have with him an outfit, comprising bed, bedding, victuals, cooking paraphernalia, and *insect-powder*. It sounds rather forbidding on paper, but labour is cheap, and a little experience enables one to keep the size of outfit within reasonable limits. The necessary coolies should always be obtained through a respectable agency and an agreement made in writing, stating all necessary details. A head-man, called a " Fu-tou," should be given charge of the coolies.

In parts of China where foreigners are well known, it is possible to dispense with the luxury of a sedan-chair, but it must be remembered that a sedan-chair is an outward and visible sign of respectability. It is the recognized medium of travel, and, quite apart from its real use, it is a necessity, since its presence ensures respect. In the out-of-the-way parts of China, even though it is carried piecemeal, a chair is of greater service and value to the traveller than a passport. According to treaty, all foreigners travelling in China must furnish themselves with a passport, which must be shown on request. This is a matter of considerable importance, and should never be omitted.

One thing more is necessary ere the caravan is fully equipped, and that is a good cook. Unless the traveller speaks Chinese he must have a servant able to speak broken English. A good travelling servant is hard to find, but the last thing the average traveller should dream of doing is to engage an interpreter. A good domestic servant will fill this function in so far as it is necessary.

CHAPTER IV

IN QUEST OF FLOWERS

A JOURNEY IN NORTH-WESTERN HUPEH

ON 4th June 1910 I left Ichang for Chengtu, via a new route through the wilds of north-west Hupeh. With 600 miles of overland travel ahead the caravan had been fitted up with all the skill at my command, and with enthusiasm to spur us on I felt that the difficulties would not prove insurmountable. Nearly all the men had been associated with me on former journeys of a similar nature.

We took the lesser road by way of San-yu-tung glen for Hsingshan Hsien, in consequence of the main road being congested by coolies engaged in blazing a trail for the Hankow-Szechuan Railway. The caravan consisted of twenty carrying-coolies, several men for collecting and general work *en route*, a chair for the Boy, and another for myself. My own start was not propitious. I was riding in my chair and had scarcely cleared the precincts of the foreign settlement when one of the poles snapped. This occasioned an hour's delay, but happening where it did new poles were secured without difficulty. It is never easy to make an early start the first day, and it is always advisable to count on a short stage. It was one o'clock when we reached the mouth of the San-yu-tung glen, 5 miles above Ichang, and overtook the main caravan. The weather was hot, and we only did another 15 li[1] to Sha-lao-che, making 35 li in all. This little hamlet consists of a few scattered houses, and we availed ourselves of the largest, which happened to be a wine-distillery, and the smell of stale brewing was very strong.

The journey up the San-yu-tung glen was very interesting,

[1] Ten li=three English miles.

much of the scenery being rugged and grand. The cliffs of hard limestone are usually 500 feet or more sheer, and are the home of Goral and other animals, and also of many cliff-loving plants. In the crevices and niches the Chinese Primrose (*Primula sinensis*) finds its home, but the flowers were past and the flower-stems all bent towards the cliffs to ensure the seeds being deposited in the rock crevices. This plant is the parent of our greenhouse Primulas, and in February and early March the cliffs present a wonderful picture, being covered with colonies of plants, one mass of warm mauve-pink flowers. Wherever the cliffs are not absolutely sheer, vegetation is rampant. Pine trees (*Pinus Massoniana*) fringe the summits and *Rosa microcarpa* was in full flower, otherwise there was very little blossom to be seen. Most of the shrubs being spring-flowering were in young fruit.

There was considerable delay in starting the next morning. One or two of the coolies gave up, and others had to be found. The road was vile all day, and it took us 10½ hours to cover 45 li. For the first 10 li the road continues to ascend the glen, which narrows and presents even finer scenery than that of yesterday. We passed a lovely natural grotto full of stalagmites inside, and with the dripping external rocks one mass of Maiden-hair Fern. These rocks are known throughout the Lower Yangtsze Valley as " Ichang Fern-stones," and command a ready sale.

In the glen *Parthenocissus Henryana* is abundant ; in the juvenile stage the leaves of this plant have prominent white veins, and are very attractive, but in the adult stage this variegation is lost, and they become very ordinary by comparison.

The glen soon became impassable, and we climbed the cliffs and ultimately overlooked the country generally. Terraced fields are much in evidence, and every available inch of country is under cultivation. Wheat, barley, and peas, all ripe, were the principal crops, and their yellow culms enlivened the landscape. We saw a small patch or two of the Opium Poppy hidden away under trees and of very poor quality. Pear and Plum trees are commonly cultivated hereabouts, Bamboo groves and Cypress trees abound. Here

and there we caught an occasional glimpse of the white-tailed
Paradise Fly-catcher (*Tchitrea incei*). Pheasants were calling,
and likewise the English Cuckoo.

Around Niu Ping (Cow-flat), which was our destination
for the day, much rice is cultivated, and the farmers were
busy transplanting the tiny rice-plants. The whole country
is finely terraced and is backed by limestone cliffs of Cambrian-
Ordovician Age. Near our destination we passed a fine Ginkgo
tree showing curious root-like protuberances on the branches.
In rocky places by the wayside, and especially in the walls of
the terraced fields, *Rehmannia angulata* abounds. Plants 1½
to 2 feet high carry six to a dozen large, rosy-pink, foxglove-
like flowers. The local name is " Fêng-tang Hwa " (Honey-
bee Flower).

" Cow-flat " is a tiny place of about a dozen houses. Our
quarters were cramped but comfortable, and the people very
nice. There is a road from this hamlet to Nanto, distant
30 li. When I first visited this place in 1901 I was an object
of great curiosity from the moment of my arrival to the time
of departure. I have been here several times since and am
now treated as an old-time acquaintance.

It was quite cool during the night, and a blanket was
required. At Ichang the very thought of a blanket was
enough to bring forth perspiration! We left about 6 a.m., and
after ascending and descending a series of lateral spurs finally
reached the small river which enters the Yangtsze at Nanto.
After ascending this river for a few miles we commenced a
steep ascent. Now by an easy and then by a heavy grade the
road winds in and out among the mountains, and we did not
reach our halting-place for the night until 6.30 p.m. The last
coolie arrived an hour later. The length of the whole journey
is supposed to be only 60 li, but we all agreed that it is a good 70.
Whatever the distance, it is certainly a hard day's travel.

The mountain sides are very steep, with razor-like ridges.
Terraced cultivation is everywhere carried out, rice is cultivated in the bottom-lands and maize on the slopes, with
occasional patches of Irish potato. Where it is too steep, or
for other reasons unsuitable for cultivation, the mountain-sides
are covered with shrubs and trees, chiefly scrub Oak and the

HIGHWAY LEAD THROUGH A NATURAL TUNNEL

common Pine. Small trees of *Cornus Wilsoniana* in full flower
were common here and there. Odd trees of *C. kousa*, also, in
full flower, were conspicuous in the outskirts of the woods and
copses. This small tree is exceedingly floriferous. In habit it
is flat-topped with horizontally-spreading branches, and the
flowers borne erect, well above the foliage. The white bracts,
which are so conspicuous, frequently exceed 5 inches in
diameter ; with age they become tinged with pink. The fruit is
large, orange-red, and edible. This Chinese form will probably
prove a better plant under cultivation than the Japanese form
with which gardeners are familiar. The plant loves a sunny,
well-drained situation. But the display of the day was made
by the wild Roses. By the side of streams the Rambler Rose
(*Rosa multiflora*), with both white and pink flowers, was abundant. In the woods higher up the Musk Rose (*R. moschata*)
filled the air with its soft fragrance. Here and there occurred
Actinidia chinensis, scaling tall trees and wreathing them
with white and buff-yellow fragrant flowers. In the forenoon
noted *Rehmannia angulata*, especially common on steep stony
places in full sun.

Our halting-place, Lao-mu-chia, is about 3500 feet altitude,
and consists of about six houses and a tile-factory. Hereabouts
much charcoal is burnt for export to Nanto and down river.
During the day's journey we met several men laden with bales
of Pear and Crab-apple leaves. These leaves are commonly
used as a substitute for tea, and there is a considerable export
from these parts to Shasi.

On leaving Lao-mu-chia we immediately commenced the
steep ascent of the Hsan-lung shan, and a climb of 1000 feet
brought us to the summit, where there is a small temple in a
ruinous condition. After a precipitous descent of a few
hundred feet the road meanders over and among the tops of
hills, composed of granitic-gneiss, which is rapidly disintegrat-
ing, and ultimately descends to the bed of a torrent and joins
the main road from Ichang to Hsingshan Hsien.

Near the summit of Hsan-lung shan, which is composed
of Cambrian-Ordovician limestones, the Chinese Tulip tree
(*Liriodendron chinense*) is common in the woods, and so is
Viburnum tomentosum with its sprays of snow-white flowers.

Styrax Hemsleyanum and *Amelanchier asiatica*, var. *sinica*, the June berry, are other trees with white flowers remarkable for their beauty and abundance of blossom. On the more open slopes *Symplocos cratægoides, Lonicera Maackii*, var. *podocarpa, Diervilla japonica*, and *Cratægus cuneata* made a fine display. Thin woods of *Pinus Massoniana* and Sweet Chestnut (*Castanea*) also occur ; the Pine trunks are gashed for the ultimate purpose of producing kindling wood. In open places *Rubus corchorifolius* abounds, and its red, raspberry-like fruits with their delicious vinous flavour were good eating. In the descent *Dipteronia sinensis*, a small bushy tree with erect trusses of small white flowers, occurs, and *Actinidia chinensis* is common. The hermaphrodite and male forms of this climber have large white flowers quickly changing to buff-yellow, and the fragrance is very pleasing. A form with purely female flowers is unknown. At the foot of the descent we joined the main road from Ichang to Hsingshan Hsien, and following this route we reached Shui-yueh-tsze, a village of 100 houses, situated in a tiny rice flat, at five o'clock. The people were very inquisitive, and I held an impromptu reception until bedtime.

On joining the main road, we saw evidences of the survey for the Hankow-Szechuan Railway. The proposed route was marked by bamboo poles, and on the rocks with Arabic numerals and initials in Roman letters.. The route descends a stream, just before reaching Shui-yueh-tsze, to Liang-ho-kou, and then continues down the Hsingshan River to the Yangtsze, which it connects with at Hsiang-che. Its construction even in this region promises to be a difficult task, and will call for great ability on the part of the engineers. Much tunnelling and blasting will be necessary, yet from Hankow to this point the task is simple compared with that which lies beyond. The cost will be enormous even in a land of cheap labour. It is highly improbable that the gentry, who are so violently opposed to the employment of foreign capital in this venture, realize the magnitude of the task and its ultimate cost.

The next day's journey proved interesting but arduous. By an undulating path we reached the top of the ridge, which is known as T'an-shu-ya (Lime tree Pass), from a gigantic Linden which occurs there. This tree (*Tilia Henryana*) is about 80 feet tall and 27 feet in girth, and though hollow appears to be in good health. The young leaves are silvery, and the tree, from its size, is a conspicuous object for miles around. Descending through a cultivated area we entered a glen which we followed for 20 li : the scenery in the lower end is magnificent. Cliffs of hard limestone rear themselves almost perpendicularly some 2000 feet and more. In the upper part of the glen *Pterocarya hupehensis* is common alongside the burn. An odd tree or two of the rare *Pteroceltis Tatarinowii* also occurs here. Throughout the glen Lady Banks's rose (*Rosa Banksiæ*) is especially abundant. Bushes 10 to 20 feet high and more through them were one mass of fragrant white flowers. It occurs in thousands and is particularly happy, growing on rocks and over boulders by the side of streams. *Cæsalpinia sepiaria*, with erect thyrsoid panicles of fragrant yellow flowers, is also abundant hereabouts. Growing on the cliffs, *Illicium Henryi*, with its dull crimson flowers, is also worthy of note. On issuing from the glen we struck a shallow, rock-strewn stream of considerable width, and after ascending it for a short distance made a very precipitous ascent of a couple of thousand feet. Crossing over a ridge and a flat area, a descending road led to Shih-tsao-che, which we reached as night was closing in. This hamlet consists of about a dozen houses scattered through a narrow valley.

During the day I collected specimens of thirty different kinds of woody plants. The striking plants of the afternoon's journey were the Amelanchier and *Dipelta floribunda*, both masses of flower. Walnut (*Juglans regia*) and Varnish trees are abundant above 3000 feet ; the sides and tops of the mountains are clothed with woods of Oak and Pine, particularly the former. We also saw many fine Willow and Ailanthus trees. *Primula obconica, Lysimachia crispidens*, and a blue-flowered Salvia are abundant up to 2000 feet. Near the inn a few trees of *Catalpa Fargesii* occur, but were not yet in flower. Hereabouts *Daphne genkwa* is abundant, but it was scarcely in flower at this altitude.

It rained a little in the early morning and showers fell at intervals during the day, nevertheless, the weather was good for travelling, since it was not too hot. Most of the journey was downhill. Soon after starting in the morning we crossed one or two low ridges, intercepted by narrow plateaux, and about noon commenced the descent to Hsingshan Hsien. The descent is precipitous in parts, but the mountain-sides are mostly under cultivation. About half-way down coal is mined, but the quality appears to be indifferent. Lime is burnt in small quantities and paper-mills occur near Hsing-shan.

Hsingshan, the only district city in these wilds, may claim to be one of the smallest and poorest Hsiens (*i.e.* cities of the fourth class) in the whole of China. It is situated on the left bank of a stream and contains scarcely a hundred houses, most of which are in a ruinous state. The wall facing the river varies from 4 to 12 feet in height. A road, apparently the main road, runs along the top of this wall. The east gate is closed by sewage ; the north gate is so low that one has to bend the head when passing through ! The whole town is dull and lifeless, as far as business is concerned, but children are plentiful, as they are everywhere else in China. The town is backed by a steep mountain, up two sides of which a wall is carried : most of the mountain-side enclosed within the wall is given over to terraced fields. The river is broad, with a shingly bottom, and the water clear and limpid. Thick-bottomed boats ply between Hsiang-t'an and Hsiang-che, a village at the head of the Mitan Gorge, on the Yangtsze. No one stays in Hsingshan, and we journeyed on to Hsiang-t'an. This name signifies " fragrant rapid " : the waters may perhaps be sweet, but the village is foul and stinking. We had some little difficulty in securing lodgings, poor as they were, and an objectionable coolie had to be evicted before we could settle down for the night.

Flowers were not common during the day. We passed a magnificent tree of *Keteleeria Davidiana*, 80 feet tall and 16 feet in girth. This tree shelters some graves, and was probably planted long ago. In the descent we passed through orchards of *Cratægus hupehensis*, all in full flower. This Hawthorn is one of several kinds cultivated in China for their edible fruit. The interesting *Torricellia angulata* occurs sparingly, and here and there large plants of *Mucuna sempervirens* cover large trees. *Catalpa ovata* is common on the plateaux and an interesting small-leaved Poplar occurs around farmhouses, but is rare.

Hsiang-t'an being in water communication with the Yang-tsze boasts quite a considerable trade. Medicines are the principal export. Rifle-stocks, roughly shaped out of Walnut wood, are exported from this neighbourhood to Hangyang in increasing quantities annually. They are worth locally 300 cash (about 6d.) each. The village is situated on the left bank of the river, and possesses an Opium Likin and a Viceroy's Bank. Pigs seem more in evidence than human beings, as judged from the four visits I have paid the place in different years. Being only 300 feet above Ichang, Hsiang-t'an enjoys a hot, dry climate.

Leaving Hsiang-t'an we immediately crossed the river by ferry and ascended a narrow valley, which soon becomes a ravine and finally a wild, entrancing gorge. At the head of this gorge we took a small mountain-path which entailed a severe climb from the river-bed to the tops of the surrounding mountains. In this ascent the Musk Rose was a wonderful sight, and *Loropetalum chinense* abundant but out of flower. Once on top of the mountains an undulating path leads to Peh-yang-tsai, where we found lodgings in a new and fairly clean farmhouse.

In the gorge I gathered *Rehmannia Henryi*, a herb less than 1 foot tall, with large, white, foxglove-like flowers. Here-abouts the root-bark of Lady Banks's rose is collected, and after being dried is pressed into bales for export to Shasi. This bark is used for dyeing and strengthening fish-nets, and it is claimed that it renders the net invisible to fish. In the valley *Kœlreuteria bipinnata* occurs, but is rare ; the flora of the ravine generally is similar to that of the San-yu-tung glen.

The mountains are clad with Oak (largely scrub), *Pinus Massoniana*, and Cypress. A few Keteleeria trees occur and also *Liquidambar formosana*. *Populus Silvestrii*, with its light grey bark, is a very common tree hereabouts. Wood Oil trees were a wonderful sight and most abundant. In the ravine they were in full leaf, and the fruits were swelling, but from 1500 feet to 3000 feet they were leafless and covered with flowers.

By the side of streams at low altitudes the Rambler Rose (*Rosa multiflora*) was a pretty sight with its white and pink blossoms, but the Musk Rose (*R. moschata*) was the flower of the day—bushes 6 to 20 feet tall and more in diameter, nothing but clusters of white fragrant flowers. Growing on some old graves I found a sulphur-yellow flowered form of *Rosa Banksiæ*; this, I think, must have been planted. Rose bushes are a special feature in this region and numerically are the commonest of shrubs. Around our lodgings the Hardy Rubber tree (*Eucommia ulmoides*) is cultivated for its bark, which is valued as a tonic medicine.

Peh-yang-tsai is a scattered hamlet, situated in a narrow valley, some 2500 feet altitude. Facing our lodgings is a massive peak called Wan-tiao shan, its face a sheer precipice of hard limestone, the summit and farther slopes apparently well forested. The people of this hamlet, like the country people everywhere in these parts, were extremely nice and obliging, and it was a real pleasure to be amongst them.

Wan-tiao shan looked too tempting to be passed by without investigation, so we spent a day, and a very hard day too, in making its ascent and descent. Leaving our lodgings at 8 a.m., several hours were occupied in rounding the spurs and surmounting the cultivated and scrub-clad land which subtend the mountain proper. At 6000 feet we reached Bamboo scrub, and a path through this led to an area where medicinal Rhubarb was cultivated, and where the drug " Tang-shên " was extraordinarily abundant. At 6500 feet we entered the timber. At the margin of the woods, to the left of the road, are extensive plantations of the drug " Huang-lien." This interesting plant (*Coptis chinensis*) is grown under a framework of brushwood reared some 3 to 4 feet above the ground. The drug is used as a tonic and blood-purifier.

As the path winds the trees are at first small, with plenty of Bamboo scrub, but this belt is very narrow and speedily gives place to large trees which extend to within 500 feet of the summit, where Bamboo scrub again becomes troublesome. Everywhere above 5000 feet, where the woods are thin and sunlight penetrates freely, Bamboo scrub is found, rendering travel excessively arduous and, unless a path is cut, im-

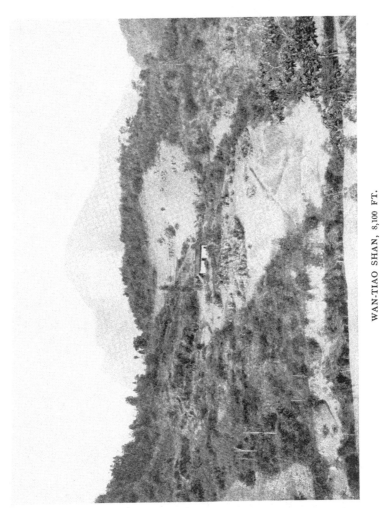

WAN-TIAO SHAN, 8,100 FT.

possible. In the dense shade of the forest the Bamboo does not thrive.

The forest, though full of splendid timber, is not rich in variety. The Chinese Beech (*Fagus sinenis*) is the commonest tree. This species always has many trunks, and trees 60 to 70 feet high, with stems 3 to 6 feet in girth, abound. The interesting *Tetracentron sinense* is very abundant ; trees 60 to 70 feet by 8 to 10 feet girth are plentiful. The leafage of this tree is very thin and characteristic. Large trees of White Birch and of several species of Maple occur scattered through the forest. The smooth-leaved Davidia (*D. involucrata*, var. *Vilmoriniana*) occurs sparingly, and good-sized trees of various Cherries, Bird Cherries, Mountain Ash, and Wild Pear are common. Rambling over the tops of the largest trees is *Berchemia Giraldiana*. Several species of Rhododendron occur ; one species (*R. sutchuenense*) forms a tree 30 feet and more tall and 5 feet in girth. Shrubs in variety abound ; in the glades *Viburnum tomentosum* was wreathed in snow-white flowers. In more open places the Musk Rose is rampant, and near the summit *Rosa sericea* is abundant.

The summit forms a sloping, undulating flat, about an acre in extent, covered with grass and a few shrubs. On the apex stands a small temple now partly in ruins. A sharp, rocky ridge extends from the summit, linking the mountain up with the ranges to the northward. The face on two sides is a vertical precipice, 2000 feet and more sheer. From the summit (alt. 7850 feet) we got an extensive view of the surrounding country. Nothing but mountains on every side ; to the north and north-west these are heaped one beyond another in quick succession and are separated by narrow defiles down which torrents rush and roar. Very difficult looked the country in front of us, but the call of the unknown was strong. We descended by the same devious path, indeed, there is no other, and reached our lodgings as darkness overtook us. Specimens of some forty odd different plants rewarded the day's labour, several of them new and uncommonly interesting. On the extreme summit Box is a common shrub, and growing with it I discovered a new species of Lilac (*Syringa verrucosa*). The following day we continued our journey northwards.

Just beyond Peh-yang-tsai we passed through copses of small Oak (*Quercus variabilis*), where the Jew's ear Fungus is cultivated. The culture is as follows : Oak saplings, about 6 inches thick, are cut down, trimmed of their branches, and cut into staves 8 to 10 feet long. These are allowed to lie on the ground for several months, where they become infested with the mycelium of the fungus. They are then stacked slantingly in scores or thereabouts, and the fructifications of the fungus develop. These are ear-shaped and gelatinous and are by the Chinese esteemed a delicacy. I tried them, but did not find them very palatable, and the experiment resulted in a bad stomach-ache !

On leaving these plantations the road descends to a ravine along which it meanders for a mile or two. Many shrubs were in flower in the ravine, and I gathered amongst other things specimens of a new genus, allied to Holbœllia, with fragrant yellow flowers. (I subsequently secured seeds of this plant, since named *Sargentodoxa cuneata*, and succeeded in introducing it into cultivation.) At the head of this ravine a steep ascent through woods of Oak and Birch leads to a cultivated area where there are two or three scattered houses and many Tea bushes. Near one house the Chinese Coffee tree (*Gymnocladus chinensis*) occurs ; the pods of this tree are saponaceous and are esteemed for laundry purposes.

From the Tea plantations the road leads through Pine woods, now by an easy, now by a heavy grade, but always ascending, and we were all glad when our destination (Hsin-tientsze) was reached. Near this place are some fine old woods, rich in a variety of deciduous trees and shrubs. I noted a Horse Chestnut (*Æsculus Wilsonii*), two kinds of Beech, *Styrax Hemsleyanum, Meliosma Veitchiorum*, the Davidia, and many different kinds of Maple and Oak—all of them large trees. In the margins of the woods *Viburnum ichangense* was particularly fine, and many Cherry trees, with both pink and white flowers, common. In moist shady places in the woods a blue Primrose (*P. ovalifolia*) carpets the ground for miles. The yellow-flowered *Stylophorum japonicum*, an Epimedium, and various species of Corydalis are abundant in and near the woods.

The hamlet of Hsin-tientsze, alt. 5600 feet, consists of one rather large house.   It is built on a slope a few hundred feet below the summit of the ridge, and from the front of the house a wonderful view of the surrounding country is afforded. Nothing but mountains as far as the eye can range, and not 20 square yards of level ground in sight! Our quarters, though cramped, were, all things considered, fairly comfortable, and were as good as could be expected in these wilds.

The next morning we made an early start in order to cover the 60 li between Hsin-tientsze and Mao-fu-lien.   Immediately on leaving we traversed an old wood especially rich in species of Maple.   Davidia and Beech are also common, whilst the interesting *Cornus sinensis* occurs sparingly as a thin tree 60 feet tall.   *Pinus Armandi* is present, but Conifers generally are very scarce in this particular locality.

We meandered around the mountain-sides, by a tortuous ascending path, until we reached a gap in the ridge and crossing over made a breakneck descent of a couple of thousand feet. A new kind of Poplar, having the young foliage bronzy-red, was common on all sides, and in the descent I gathered *Primula violodora, Rhododendron Augustinii, Acer griseum*, and a pink-flowered Staphylea, the last two both small trees.   The most interesting find, however, was a new Hydrangea (*H. Sargentiana*), a shrub 5 to 6 feet tall, with stems densely felted with short bristly hairs and large, dark green leaves with a velvety lustre— in foliage alone this species is strikingly handsome.

At the foot of the descent we came upon small woods of *Pinus Henryi*, a tree averaging 60 feet in height, more or less pyramidal in shape, with bark usually rough and black, but sometimes red in the upper parts.   The cones vary considerably in size and are retained on the tree for several years.   In the valley near the Pine woods there is considerable cultivation. Walnut trees are common and Cunninghamia abounds.

Leaving this valley, a long but fairly easy ascent led to the top of another ridge, and a precipitous descent brought us to another narrow valley.   These ascents and descents were most fatiguing and occurred with exasperating frequency every day, and several times a day at that.   Another climb of over 2000 feet and we reached our destination for the day, finding accommodation in an inn which is also a large medicine depot, and is owned by a wealthy man from the province of Kiangsi. This inn is a large, rambling two-storied structure with several outhouses and a large courtyard. There is not sufficient level space to accommodate the whole place, and the front part is supported on posts. It serves as general store for the whole country-side, and in addition is a veritable museum. Dirt in every shape and form draped everything, and the stink from adjacent piggeries was tempered with the odour of various aromatic herbs. The business instinct of the house is strong, as I found to my cost when changing some silver and buying a goat. The rites of ancestor-worship were strictly carried out every morning and evening, and everything done to ensure continued and increasing prosperity. The burning of incense and candles and the performance of mystical genuflexions may assist business, but a little more attention to cleanliness and sanitation would make a stronger appeal to the foreigner. At least, such were my conclusions after a thirty-six hours' stay in the place.

It rained a good part of the next day, but as we had decided upon a day's rest it did not inconvenience us. In the forenoon I went out for a few hours to investigate the woods around Mao-fu-lien. Some very large trees of Sassafras (*S. tzumu*) occur here—the largest specimen is nearly 100 feet tall and 12 feet girth. The Chinese Sassafras has no medicinal value, and the wood is used for box-making and fuel only. Oak and Sweet Chestnut are plentiful and form small woods. The Chestnut (*Castanea Vilmoriniana*) is a singular species, with a single ovoid nut inside the spiny fruit ; the flowers have a peculiarly unpleasant smell. Around the inn are cultivated many trees of the Hardy Rubber and also *Magnolia officinalis*. Walnut and Varnish trees are abundant, and behind the house is a fine flat-leaved Spruce (*Picea pachyclada*). The mountain-tops are clothed with Grass, Brambles, scrub Oak, bushes of the pink-flowered *Rhododendron Mariesii*, and the scarlet *R. indicum*.

The view from the inn is one of steep ridges and high mountains, separated by deep, narrow chasms as far as the eye can range. It is indeed a fascinating country, but exhausting to travel over.

IN SAN-YU-TUNG GLEN, CUPRESSUS FUNEBRIS IN FOREGROUND

A fine morning followed yesterday's rain, the country looked refreshed, and the air was laden with fragrance from the myriad flowers on every side. The coolies grumbled loudly over the extortionate charges at the inn, and several hours elapsed before they recovered their cheerfulness. The day's journey commenced in a steady ascent to the top of a ridge followed by the usual precipitous descent. Hereabouts *Staphylea holocarpa*, a small, very floriferous tree, with both white and pink flowers, is very common and most strikingly beautiful. Another interesting plant is *Salix Fargesii*, a dwarf-growing Willow, having large very dark green leaves. A small torrent marks the foot of the descent, and from this point on we occupied several hours in an exhausting climb to the summit of another ridge, finally crossing over at 7300 feet altitude. In the ascent a new Spruce, having short square leaves and small cones, was discovered, and many small trees of Hemlock Spruce were noted. Near the head of the ridge, on cliffs, Box (*Buxus microphylla*) is very common, and a rosy-red flowered Primrose is abundant in grassy places. A dwarf Bamboo forms dense thickets on the top of the wind-swept ridge.

The descent quickly leads into copses of Birch, and later into fine woods composed of mixed deciduous trees and shrubs and a few conifers. In these woods we spent a profitable time, collecting in all specimens of some fifty different kinds of woody plants. We saw one or two large trees of Davidia and many of Tetracentron. Cherries in variety are plentiful, and were a wonderful sight—nothing but masses of pink and white. Three kinds of Rhododendron were collected, and six in all noted. Maples in variety are very common, but one large tree of *Acer griseum*, with its chestnut-red bark, exfoliating like that of the River Birch, was the gem of all. Various *Pomaceæ* and one or two species of *Lauraceæ* make up a fair percentage of the small trees. Viburnums in variety, Honeysuckles, Diervillas, Deutzias, Philadelphus, and *Neillia sinensis* are everywhere abundant. In rocky, more open places *Viburnum rhytidophyllum* with its long, thick wrinkled leaves looked particularly happy, and in places exposed to the sun a Crab Apple (*Malus* sp.) with pink flowers was a sight for the gods. On wet, humus-clad rocks *Pleione Henryi* luxuriates, and herbs in endless variety crowd every available spot. A fine torrent collects up the waters of countless smaller streams, and falls down the narrow ravine, often in a series of waterfalls hundreds of feet high, the noise of the falling water alone breaking the silence of the forest depths.

With some difficulty, owing to the timidity of the people, we obtained lodgings in a peasant's hut at Wên-tsao, alt. 6150 feet. This tiny hamlet consists of four small houses, scattered and pitched on the steep mountain-slope. It is surrounded on all sides by precipitous mountains covered with forests. Around the houses small patches have been cleared, and wheat, a little maize, and a few peas and vegetables are cultivated.

The forests of this region are particularly rich, and in order to better appreciate them I propose to interpolate here extracts from my journal of another date :—

" *May* 30.—Wên-tsao. On a precipitous slope facing our lodgings a score or more Davidia trees occur ; they are one mass of white, and are most conspicuous as the shades of night close in. Two large trees of *Pterostyrax hispidus* are growing amongst these Davidias, and are laden with pendulous chains of creamy-white flowers."

" *May* 31.—Go over and investigate the Davidia trees and the forests generally. Crossing a narrow neck a woodcutter's circuitous path leads us down to a narrow defile through a fine shady wood. Ascending a precipice with difficulty, we soon reach the Davidia trees. There are over a score of them growing on a steep, rocky declivity ; they vary from 35 to 60 feet in height, and the largest is 6 feet in girth. Being in a dense wood they are bare of branches for half their height, but their presence is readily detected by the numerous white bracts which have fallen and lie strewn over the ground. The tree starts up from below when felled ; indeed, it naturally throws up small stems after it gets old. The bark is dark and scales off in small, irregular flakes. By climbing a large Tetracentron tree growing on the edge of a cliff, and chopping off some branches to make a clear space, I manage to take some snapshots of the upper part of the Davidia tree in full flower. A difficult task and highly dangerous. Three of us climb the tree to different heights and haul up axe and camera from one to another by means of a rope. The wood of Tetracentron is brittle, and the knowledge of this does not add to one's peace of mind when sitting astride a branch about 4 inches thick with a sheer drop of a couple of hundred feet beneath. However, all went well, and we drank in the beauties of this extraordinary tree. The distinctive beauty of Davidia is in the two snow-white connate bracts which subtend the flower proper. These are always unequal in size, the larger usually 6 inches long by 3 inches broad, and the smaller 3½ inches by 2½ inches ; they range up to 8 inches by 4 inches and 5 inches by 3 inches. At first greenish, they become pure white as the flowers mature and change to brown with age. The flowers and their attendant bracts are pendulous on fairly long stalks, and when stirred by the slightest breeze they resemble huge Butterflies hovering amongst the trees. The bracts are somewhat boat-shaped and flimsy in texture, and the leaves often hide them considerably, but so freely are they borne that the tree looks, from a short distance, as if flecked with snow. On dull days and in the early morning and evening the bracts are most conspicuous. The fruit superficially resembles a small walnut, but the inner shell is absolutely unbreakable. To my mind *Davidia involucrata* is at once the most interesting and beautiful of all trees of the north-temperate flora.

" With the Davidia is a good-sized tree of the Horse Chestnut (50 feet in height by 4 feet in girth). Higher up Hornbeam and Tetracentron are common, and Birch, white, red, and black, luxuriate.

" Maples are a feature of these woods ; all are tall trees, but of no great thickness. Unfortunately very few are flowering, and indeed this is true of the forest trees generally this year.

" Perhaps the commonest tree in these forests is the Beech ; parts being formed entirely of these trees. So light-demanding are they that they suffer no competitors or even undergrowth. For the first time it is possible for me to say definitely that two distinct species of Beech exist in this region. One forms a tree with a single trunk, the other always has several trunks. The former species has glabrous, shining green leaves, a large,

dense, much-branched head ; it makes a tree 40 to 50 feet
high with a trunk 5 to 10 feet in girth, and, save for its smaller
stature, very strongly resembles the European Beech. The
second species, which is the recognized Chinese Beech, grows
much taller, but never attains the girth of the other. It
generally has six to twelve trunks, averaging 2 to 5 feet in girth,
arising closely together and slanting away from one another
as they grow. The bark is light grey and the leaves sub-
glaucous and hairy below ; branches somewhat ascending
but with the young branchlets slender and pendulous. A
local name for the Beech is ' Peh Litzu.' Small plants are
common, but no flowers are to be discovered.[1]

" In the shade of trees, *Ribes longeracemosum*, var. *Wilsonii*,
a remarkable black currant, with racemes 1 to 1½ feet long, is
common, whilst *Rodgersia æsculifolia*, with large, erect, thyrsoid
panicles of white flowers, is rampant.

" Five species of Oak—three deciduous and two evergreen
occur. *Meliosma Veitchiorum* and many species of *Pomaceæ*
and Cherries are common, whilst the Varnish tree is everywhere abundant. In dense shade various evergreen Barberries
occur, and in open country *Neillia sinensis* forms dense thickets.

" Of Conifers, *Pinus Armandi* and *P. Henryi* are scattered
over the cliffs ; *Picea Wilsonii* and a flat-leaved Spruce (*P.
pachyclada*) are rare, whilst the Hemlock Spruce [2] is fairly
common on the cliffs—neat, dense trees of no great size with
their young leaves just unfolding and old cones abundant.
The White Pine (*P. Armandi*) is more common higher up on
the mountains ; with its long needles, graceful port, and light
grey bark this tree is strikingly handsome ; the cones are
pendulous, borne at the ends of the glabrous branches. The
very resinous wood is used locally for torches, burning with a
clear, bright flame, and gives a good light."

[1] In 1910 I succeeded in introducing young plants of both species into
the Arnold Arboretum from this region.
[2] *Tsuga chinensis*.

CHAPTER V

FOREST AND CRAG

ACROSS THE HUPEH-SZECHUAN FRONTIER

ON leaving Wên-tsao a sharp descent for a couple of hours brought us to the upper waters of the Hsingshan River, which we left several days ago. Crossing this stream by a covered bridge we reached the hamlet of Li-erh-kou. Around this hamlet trees of the Hardy Rubber (*Eucommia*) and *Magnolia officinalis* (Hou-p'o) are cultivated for their bark. A steady ascent from Li-erh-kou through occasional woods of Oak and Birch, interrupted by areas where people were busy ploughing the fields and sowing maize, brought us to the hamlet of Chintien-po, where we lunched. Near this place is a fine new Meliosma (*M. Beaniana*), a tree 60 feet high. It was leafless, but one mass of creamy-white flowers borne in pendulous panicles. Near by this tree I discovered one small specimen of the " Judas tree " (*Cercis racemosa*). Prior to this discovery I knew of only two trees some fifteen days' journey south-west of Ichang. This new tree was about 25 feet high, with a stem half decayed through at the base, and a mop-like head. In spite of its partial decay the tree appeared in vigorous health, and was one mass of silvery-rose coloured flowers, borne in short racemes. The leaves of this species are hairy below. Varnish and Walnut trees occur in abundance, and we met several coolies laden with cakes of fat, expressed from the fruits of the Varnish tree (*Rhus verniciflua*). The double-flowered form of *Spiræa prunifolia* is commonly planted on graves, and the bushes were wreathed in flowers.

Soon after leaving Chin-tien-po we commenced a precipitous ascent, and after climbing for several miles reached the neck of a ridge where *Viburnum rhytidophyllum* luxuriates. From this neck the ascent is more gradual, and but few crops are grown, as it is nearing the limits of cultivation in these regions. Near some limestone cliffs are two magnificent trees of Maackia, each 60 feet tall and 7 feet in girth. The bark of this tree is smooth, of a light grey colour, and the unfolding leaves are silvery grey. Here, too, are many small trees of the Bladder tree (*Staphylea holocarpa*) and Peach bushes. These were in full flower, and flitting amongst the flowers and drinking in the honey were many beautiful little sun-birds (*Æthopyga dabryi*). *Rhododendron indicum* was left behind at 5500 feet altitude.

A few hundred yards beyond the limestone cliffs we crossed over at 7000 feet altitude, into Fang Hsien, and traversed a narrow moorland valley clothed with grass and bounded by rounded hills covered with thickets. In this moorland are acres of *Astilbe Davidii* and *A. grandis*, with several Senecios and other ornamental herbs. The thickets are composed chiefly of Birch and Willow, with a few Poplar and Silver Fir, and an occasional flat-leaved Spruce. The vegetation was scarcely in leaf, and it was evident from the appearance of the ground that snow had only just melted away. We flushed a Solitary Snipe and secured a cock pheasant for the larder, but very little life of any sort was visible in these uplands. At the head of this moorland valley we entered a narrow defile and, after skirting the side of a mountain through thickets in which various Maples and Currants were prominent, reached Hung-shih-kou. This is a miserable hut of wood in a half-ruinous condition, kept by a family clothed in rags. It is situated at an altitude of 6300 feet, by the side of a considerable torrent, and is walled in by precipitous, well-wooded mountains.

At night some of the coolies slept in a loft above the room I occupied, and every movement they made caused dust and dirt to fall over my bed. On waking in the morning I found myself covered with this filth, and nearly choked with the dust into the bargain. The owner of this hovel is a hunter, and he has shot the Serow of this region, which is known as " Ming-tsen Yang." He had a couple of pairs of horns and a flat skin which we secured, and, judging from this fragmentary material, the beast must be larger than any known species of Serow. (In 1907 my associate, Mr. Zappey, made several trips after this

THE CHINESE PRIVET (LIGUSTRUM SINENSE) 20 FT. TALL

animal, but to no purpose, though he secured a tantalizing glimpse of just one specimen.) The name Hung-shih-kou signifies " Red stone mouth," and has reference to the outcropping of red sandstone which occurs here and extends to Hsao-lung-tang, 20 li distant, which we made our halting-place for the next day. Though we had only 20 li to cover we started early, glad to escape from the miserable lodgings into the woods again. Ascending a stream, through brushwood thickets composed of Willow, Birch, Spiræa, and Roses, we twice crossed the stream by rotten bridges of roughly hewn tree-logs before reaching our destination. On the way we passed several fine trees of *Picea Wilsonii*, beneath which old graves nestle. The largest trees are about 70 feet tall and 6 feet in girth ; the leaves bright green, and the habit distinctly stately ; the cones are borne in large clusters, and many still remained on the trees. Here also are small trees of the White Pine (*Pinus Armandi*) with cones 9 inches long. A new Poplar was discovered in flower, and Veitch's Viburnum and Spiræa were common with their young leaves just unfolding.

The handsomest tree in these parts is, however, the Chinese form of *Betula utilis*, a Birch with orange-red bark, which on exfoliating exposes the glaucous waxy bloom of the layer below. Trees 40 feet high are still pyramidal in habit, much branched, with slender, ascending branches on which the lenticels are very prominent. The older trees, as seen on the tops of the mountains, are mop-headed, 60 to 80 feet tall, with a clean trunk for 40 feet more, and are still strikingly handsome though blown and battered by the wind.

The hamlet of Hsao-lung-tang (Small Dragon-pool), alt. 7400 feet, consists of two dilapidated wooden huts pitched on opposite sides of a lovely burn, which flows through a narrow sloping valley lying almost due east and west. This valley is flanked by steep ridges clad only with grass and scrub. Odd patches of Birch and Silver Fir attest to forests which have all been destroyed by fire. From the numerous old graves and abandoned fields it is evident that formerly more people dwelt in this valley than do so to-day. Tiny patches of cabbage and Irish potato occur around the huts ; and also plantations of Tang-kuei (*Angelica polymorpha*, var. *sinensis*), a valued

Chinese medicine. The people declare the valley too cold for wheat or barley !

On the occasion of my first visit to this place in 1901, I had to retrace my steps owing to dearth of supplies. Since that date no white man had visited this region. In the direction in which we were bound these are the last inhabited houses for over a hundred li.

I took a photograph of the hostel on my arrival, but what I should have liked to photograph was the interior. This was impossible, since, even at midday, a light was necessary to see into the farthest corners. Dirt and filth in many forms abounded, and although plenty of timber is to be had for the felling, the house, through the idleness of its keeper, has been allowed to fall into a most ruinous state. Of one low story, the house is bisected into four compartments, and is provided with no outlet for the smoke or for the ingress of light, save through the doorway and holes in the roof ; the floor, of course, is mother earth. Pigs were quartered in one section, into which our arrival also forced the owners. Cows and goats occupied a hovel 6 feet from the door, the floor of which was fully a foot deep in filth. Luckily, the weather continued gloriously fine, and the miserable surroundings were less evident in consequence. (In passing, I might record the fact that this was the only occasion on which I enjoyed fine weather in this place. Twice previously I had been marooned here for days, and either stayed in bed or shivered by the doorway watching the rain.)

Bee-keeping is one of the principal industries of the peasants in these wilds, and around this hostel are scores of bee-hives. The hives are hollowed-out logs of Silver Fir, about 3 feet 6 inches long by 1 foot wide, two pieces of wood are fixed crosswise in the centre, and opposite these three or four holes are bored to allow the bees ingress and egress. Rude boxes often take the place of these logs. The beeswax is not separated from the honey, the honeycomb being eaten as removed from the hives. Though the climate is rigorous, the bees are healthy and strong, and disease is unknown among them.

The morning following our arrival we ascended the Shamu-jen range behind our lodgings. The first 500 feet was steep going, but afterwards the climb was easy. At about

8000 feet woods of Silver Fir occur. The trees at first are of no great size, but their dimensions increased as we ascended. Most of the larger trees have been felled and converted into coffins ; the remains of thousands of them are scattered everywhere around. On the decayed trunks of many of these trees large bushes of Rhododendron are growing, thereby proving that the trees have lain there these many years past. Some of the prostrate trunks measured over 150 feet in length and 6 feet in diameter. None of this size is now standing, but plenty that are over 100 feet tall occur. The upper part of the ridge is a cliff some 200 feet high, under the lee of which Birch and Maple are common and wild Rhubarb is also found. We discovered a more or less easy path up the cliff, and crossed over at 9700 feet altitude. The highest peak in this range is probably a couple of hundred feet higher. The summit is of hard limestone with rare outcroppings of red sandstone. Stunted wind-swept Silver Fir and various kinds of Currant extend to the summit. Rhododendron and a dwarf Juniper (*J. squamata*) are also common. The descent was through woods of Birch and Bamboo to an open, grassy, scrub-clad, sloping moorland, through which a considerable torrent flows. The Bamboo, so common hereabouts, is very beautiful, forming clumps 3 to 10 feet through. The culms are 5 to 12 feet tall, golden yellow, with dark, feathery foliage ; the young culms have broad sheathing bracts protecting the branchlets. Taken all in all, this is the handsomest Bamboo I have seen.[1]

In the vicinity of the stream shrubs in great variety abound ; of these the Willows, Roses, Spiræas, Philadelphus, Hydrangeas, odd bushes of *Rhododendron Fargesii*, and clumps of *Aralia chinensis* are the principal features. The Rhododendron referred to is one of the most beautiful, with compact trusses of white or, more commonly, rosy-red (occasionally deep red) flowers ; the leaves are small, displaying the trusses of flowers to great advantage. This species is usually a bush 5 to 8 feet tall, and of about the same dimensions through the head ; more rarely it is 15 to 20 feet tall. The steep grassy slopes are almost devoid of trees ; the fine pasture land and the typical moorland character of this narrow valley

---

[1] In 1910 I successfully introduced it into cultivation.

constitute a region that is very different from others in central China.

In the afternoon we visited Ta-lung-tang (Large Dragon-pool), a deep, silent pond about a stone's throw across, nearly circular in outline with reedy margins, walled in by steep grassy mountain slopes. In short, in situation and appearance the very kind of pool that in any country legends would be wrapped around, and so in this case many curious stories concerning elfs and demons are centred round this silent pool. The day was gloriously fine and sunny, but the wind, which swept through the valley in considerable force, was very cold. Whether it be due to local conditions or to the altitude I could not determine, but the tree flora is comparatively poor and of little interest, and very unlike the belts that occur between 4000 and 6500 feet. The altitude, however, favours coarse herbs, and these are rampant. Many interesting shrubs also occur, but with the exception of Silver Fir, Birch and Poplar trees are rare.

With a prospect of 60 li of unknown road before us we planned a daylight start, but this scheme did not mature, as the men had to prepare and cook their morning meal before starting. The entire absence of food supplies makes travelling hereabouts extraordinarily difficult. Yesterday four of the men journeyed back 45 li in order to buy food-stuffs, and returned only after dark ; several of them were up most of the night grinding maize and preparing cakes for the march.

On leaving Hsao-lung-tang we ascended the lesser branch of a stream through a narrow valley flanked by bare grassy mountains having here and there small patches of Silver Fir and Birch forest. The road is one steady climb, never steep but often difficult owing to the Bamboo scrub. The decaying stumps and stark tree trunks speak eloquently of the magnificent forests which must have formerly existed here until destroyed by axe and fire. To the botanist and lover of Nature this vandalism is painful, but presumably it was necessary for economic reasons. The unwitting cause of it all has been the Irish potato. But Nature took her revenge when, twenty-three years ago, the Potato disease devastated

CERCIDIPHYLLUM JAPONICUM VAR. SINENSE, 80 FT. TALL, GIRTH 7 FT.

the crop and ruined the country-side, causing a general exodus of all the people. Nature is fast reclaiming the whole region, but re-afforestation is a slow process. Nearing the head of the pass we entered large timber— a fragment of the virgin forest, composed exclusively of Silver Fir and Birch with a dense undergrowth of Rhododendron. The last named comprise four species—*R. Fargesii, R. maculiferum, R. sutchuenense,* and *R. adenopodum,* most of them bushes 10 to 20 feet tall, their flowers making one blaze of colour. The Silver Fir and Birch trees are of huge dimensions, but none was fruiting. On emerging from this patch of forest we entered a rolling moorland covered with Bamboo scrub which merges gradually into areas clad with the dwarf Juniper, coarse grasses, and herbs, amongst which a species of Onion was abundant. This moorland extends across the rounded saddle of the range and for several miles down the other side. The crest of the saddle I made 9500 feet altitude, and from this point we obtained a fine view of the series of bare, savagely jagged peaks from which the range (Sheng-nêng-chia) takes its name. The highest peaks probably exceed 11,000 feet altitude, and the lower slopes are forested, but the country is not attractive. Animal life is remarkable for its absence, and hardly a bird was to be seen. The solitude which reigned in this remote, inaccessible region was broken only by the noise of rushing waters and the low whining of the wind amongst the tree-tops. In shady places blocks of ice still remained, and about the head of the pass the grass was only just beginning to show green. Save for an alpine Primula and a Dandelion no flowers of any sort were to be seen.

On crossing the pass we again entered Hsingshan Hsien, and after wandering across moorland for a few miles a short steep ascent led us across a lateral spur into Patung Hsien. From this point a precipitous descent of 2000 feet brought us to a ruined and deserted hut at a place called Wapêng, the only accommodation the country-side affords. In the descent we passed hundreds of curious rock-stacks—bare blocks of shale standing erect, with acute edges, like gaunt sentinels guarding the neighbourhood. The mountain-side was formerly under cultivation, but is now abandoned and covered with grass and coarse herbs.   Around the hut a little Medicinal Rhubarb and much Tang-shên was growing, telling of former plantations of these and other medicines.   The country on all sides is very steep and much cut up, but stark decaying tree trunks, the sole remnants of former forests, mar the beauty of the landscape on all sides.

We reached Wapêng (alt. 8400 feet) fairly early in the afternoon, and the men were busy till nightfall collecting fuel and rigging up a bamboo shelter beneath which to pass the night.   The day had been gloriously fine and the night proved equally so, with a distinctly frosty nip after sundown.   A roaring fire made things look cheerful, and everybody was in the best of health and spirits.   The sides of the hut were airy and the wind played about one all night.   The roof was partially wanting and afforded a good view of the starry heavens above.   It was a lonely place, yet one felt peculiarly happy and glad to be privileged to visit a region so remote from the world in general.

There was no difficulty in getting the men up next morning, and we were off just as the sun's rays broke over the landscape. Dark mists obscured the view for an hour or so, but as the sun rose these disappeared and we enjoyed another gloriously fine day.   A steep and precipitous, nay breakneck, descent of a 1000 feet brought us to a narrow well-wooded valley, walled in by forest-clad mountains.   The Silver Fir does not descend more than 500 feet from Wapêng, below which its place is taken by Hemlock Spruce.   This Spruce is not plentiful, but giants 100 feet tall by 12 feet in girth occur.   The forests as we descended quickly become of mixed character, and finally conifers completely disappear.   The variety of trees and shrubs was astonishing, and nearly all the more interesting trees of western Hupeh were to be found and in quantity.   Maples are particularly abundant, and I gathered specimens of a dozen species in flower.   Four species of Rhododendron occur scattered, but not in quantity.   On rocks in places an interesting orchid (*Pleione Henryi*) abounds and was one mass of flowers.   The Davidia is fairly common, and the curious *Euptelea Franchetii* and *Tetracentron sinense* are the commonest of trees. A feature in these woods was *Staphylea holocarpa*, a small tree covered with pendulous trusses of white and rosy-pink flowers. A Horse Chestnut (*Æsculus Wilsonii*), the Chinese Yellow-wood (*Cladrastis sinensis*), Hemsley's Styrax and *Pterostyrax hispidus*, all of them large trees, were fairly common; Cherries, Bird Cherries and many *Pomaceæ* abound. Birch is one of the commonest constituents of these forests; in the more open areas Bamboo scrub forms dense thickets, and high up in the woods *Rhododendron maculiferum* forms trees 25 feet tall with a trunk 1 foot in diameter.

Here and there clearings have been made for the cultivation of the medicine " Huang-lien " (*Coptis chinensis*). In one abandoned clearing were hundreds of *Lilium tigrinum* luxuriating amongst the grass and tall herbs. In dark shady places the noble *Lilium mirabile* is common. This lily has tubular snow-white flowers spotted with red within, and glossy green, cordate leaves. An occasional Spruce or Pine tree occurs, and at the edge of the forests Cunninghamia appears. Many of the cliffs are clothed with Hemlock Spruce. Birch is fairly common, but, with the exception of one or two evergreen species, Oak is very scarce. Hornbeam is not plentiful, and Magnolias are decidedly rare trees; Ash is general, and the Linden, represented by three or four species, abundant. The Laurel family is represented by four species, all of them deciduous, including a handsome kind with young foliage of a bronze-red. Honeysuckles are rare, save for the climbing species *Lonicera tragophylla*, which has golden-yellow flowers. Clematis in variety are common, especially *C. montana* (white and rosy-red forms) and *C. pogonandra* with its top-shaped yellow flowers. Many species of Schisandria, all of them a wealth of flowers, *Holbœllia Fargesii*, and the botanically interesting *Sinofranchetia sinensis* are the principal climbers.

The road follows the course of a torrent which rises near Wapêng and quickly becomes a considerable stream. The path is narrow, very rocky and difficult to follow, and how our chairs got through was a puzzle. Both torrent and path ultimately plunge into a narrow ravine shut in by lofty cliffs, unclimbable and bare. In places the rocks are of limestone, but from 5000 feet downwards slate and mud-shales predominate.

At 4500 feet altitude we reached the edge of the forest and entered a cultivated area, where there are a few inhabited houses—the first we had seen for two days. Barley and Irish potato are the crops. Near the edge of the forest the torrent flows underground for about a mile. On rocks here *Lonicera pileata* abounds as a fluviatile shrub ; the curious climber *Hosiea sinensis* is common, covering rocks fully exposed to the sun. In the open country I noted in full flower a fine specimen of the Chinese Tulip tree (*Liriodendron chinense*), 70 feet tall and 5 feet in girth.

A precipitous descent, through fields margined with Tea-bushes, led to the tiny hamlet of Sha-kou-ping, where the torrent we had followed joins with a very considerable stream flowing down from the north-east. The united waters plunge at once into a ravine and finally enter the Yangtsze a few miles above the city of Patung. Sha-kou-ping is only 2600 feet above sea-level, and is hemmed in on all sides by lofty cliffs. The flora is that common to the glens and gorges around Ichang, and the wealth of flowers was extraordinary. The Banksian rose is one of the commonest shrubs hereabouts, and was laden with masses of fragrant white flowers. Opium Poppy was abundant and the whole countryside was gay with the colour of flowers. *Styrax Veitchiorum* occurs here, and trees 12 to 40 feet tall were masses of ivory-white.

From Sha-kou-ping we toiled slowly up the rocky ravine down which the main stream rushes. A paper-mill or two are located here, but houses are few and far between. The rocks are of slaty shales, often very rotten, and the torrent is a succession of rapids and cataracts. In spite of the turbulent nature of its waters it is full of fish, some of them of good size.

The hamlet of Ma-hsien-ping, our intended destination, proved to be a miserable place of some half a dozen hovels all filled with people engaged in collecting tea. We therefore journeyed on for another 10 li to some farmhouses at Shui-ting-liangtsze, and arrived just as the sun was setting

behind the range. We found accommodation in a large farmhouse, alt. 3900 feet. The day's journey proved very arduous,
but there was much by way of compensation. The scenery
was sublime and the flora wonderfully rich and varied. In all
I gathered specimens of upwards of fifty new kinds of woody
plants, many of them previously unknown. This region is one
of the richest I have visited, and I subsequently secured a fine
haul of seeds, the great majority of the plants raised from them
being now found growing and thriving in many gardens of
Europe and America. (Later I again traversed this same
region, and owing to heavy rains was over a week in crossing
the country between Hsao-lung-tang and Shui-ting-liangtsze,
a flooded torrent holding us up for three consecutive days.)

It was nearly midnight when all was quiet last night, the
men being loud in their grumbling against taking the road to
Taning Hsien instead of that to Wushan Hsien. The reports
we had heard indicated a bad time ahead for all of us and for
the men in particular, owing to the extreme poverty of the
country-side. I heard them as I lay in bed, but fortunately
no complaints were brought to me.

It was later than usual when we got away in the morning.
After a steep ascent we meandered along the mountain-side,
and ultimately crossed over into Fang Hsien again by a low
pass, alt. 5600 feet. This is the real watershed of the Han
and Yangtsze River systems. The Sheng-nêng-chia is a gigantic spur thrust out from the backbone of the chain, and the
streams which take their rise from three sides of this spur
flow down to the Yangtsze. From the watershed we had a
good view of the Sheng-nêng-chia peaks bearing E.S.E.,
and of some equally lofty mountains to the east, evidently
in the vicinity of the Yangtsze itself. On both sides of the
watershed is a rather broad cultivated valley bounded by razor-
backed hills clothed with woods of Oak and Pine. Varnish
trees abound on the edges of the fields and Walnut trees are
also common. Farmhouses are scattered over the country-side,
and the crow of the Pheasant, the coo of the Wood Pigeon
and the notes of the Cuckoo were heard on all sides. By the
wayside are many fine trees of Sweet Chestnut and Magnolia,
and one very fine specimen of *Corylus chinensis*, 120 feet tall

and 12 feet in girth. Many medicines are cultivated hereabouts, more especially Rhubarb and Tang-shên. *Populus lasiocarpa*, with huge handsome foliage, is one of the commonest trees.

After a few miles the cultivated valley ends and we entered a narrow defile flanked by steep, well-wooded mountains. Hereabouts the interesting *Sinowilsonia Henryi* is common, forming a small, bushy tree with handsome foliage and long, pendulous racemes of inconspicuous flowers. The most ornamental tree, however, is a fine Crab Apple, which was laden with umbels of pure white fragrant flowers borne on long slender stalks. Issuing from this defile we entered a small cultivated flat and found lodgings at the hamlet of Pien-chin, alt. 5200 feet.

The vegetation during the day's journey was not very remarkable, though I added sixteen kinds of plants to the collection. Noteworthy on the rocks and cliffs was *Viburnum rhytidophyllum*, with its large flat corymbs of dirty white flowers, which are not very pleasing to the nostril. In the defile the mountain-side is rich in shrubs; amongst which various Rhododendrons were prominent ; *Rhododendron indicum* was common and *Rosa sericea* was just opening its flowers. All day Oak woods were common; but these never contain much that is of more than passing interest. In abandoned cultivated areas a small Poppy, resembling the common Iceland Poppy, with deep yellow (occasionally orange) flowers was very abundant and attractive. In shady places the large yellow flowers of *Chelidonium lasiocarpum* made a fine show, and common on bare limestone cliffs are *Corydalis Wilsonii* and *C. tomentosa*, both species with yellow flowers and glaucous foliage. Around our lodgings there was much cultivation, maize, barley, pulse, and the Irish potato being the principal crops. Several papermills occur by the side of the stream, bamboo pulp being the raw material from which the paper is made.

On leaving Pien-chin we followed a river to a point where it is joined by a tributary stream which we crossed and then ascended the road which skirts its banks. This stream is gentle for a Hupeh torrent, and for 10 li the road is of the easiest. The mountain-sides are covered with shrubs and trees; among which Cercidiphyllum was conspicuous. Occasional houses and small

THE LARGE-LEAVED POPLAR (POPULUS LASIOCARPA) 50 FT. TALL,
GIRTH 5 FT.

patches of cultivation occur, but the country generally is very sparsely peopled. *Populus lasiocarpa* is abundant, and large branches are commonly driven in the ground to make fences; these branches take root and form groves. A magnificent tree of *Ailanthus Vilmoriniana*, 150 feet tall and 20 feet in girth, was passed, and I was astounded at the huge size of this specimen. Tangled masses of *Actinidia chinensis* and various kinds of wild roses were everywhere abundant, filling the air with soft fragrance. Leaving this delightful mountain stream we made a steep ascent of 900 feet and then, to our great surprise, entered a broad level valley. This valley was evidently in earlier times a mountain lake—to-day the margins are cultivated and the centre is a marsh. The whole district is known as Chu-ku-ping or Ta-chu-hu,—the latter name having reference to its former condition as a lake. A flat area of this character is unique in these regions, as far as my knowledge goes. Several roads cross this flat and we took the one for Taning Hsien. By the wayside strawberries, white and red, luxuriate and were very good eating. Quite a number of horses and cattle were grazing in this valley, and the country could support many more.

After meandering some 15 li over the easiest of roads we made a very steep and fatiguing ascent to alt. 7300 feet, and crossed over into the province of Szechuan. From the neck of the divide, looking away E.S.E., we obtained a good view of the Sheng-nêng-chia and the main and subsidiary ranges and peaks—nothing but mountains on every side save the tiny valley at our feet which we had just crossed. In the ascent we passed many shrubs in full flower; particularly striking were the various kinds of Viburnum, Deutzia, Abelia, and Cornus. A precipitous descent through a ravine and we reached the hostel at Hwa-kuo-ling, alt. 6350 feet, where plantations of Rhubarb were common and several other medicinal plants cultivated.

The road we were following is called the " Great salt road," but we only met four men carrying salt in the day's march. Indeed, on the whole journey we encountered practically no traffic. This wild mountainous country supports only a very sparse population and foreign trade has no chance hereabouts. Our great difficulty was in securing enough food for the men.

At Chu-ku-ping we managed to get one good meal from the local head-man and bought portions of a wild pig recently killed. At the hostel nothing was obtainable and the men had to eke out on the small rations they had with them. Goitre is common in these regions and nearly every one is affected. It would seem to be hereditary, since I noticed children in arms showing unmistakable swellings in the throat.

Boisterous winds and heavy clouds alternating with bright sunshine marked our first day's journey in eastern Szechuan. We were again amongst cliffs of hard limestone and the scenery strikingly resembles that of the Yangtsze Gorges and contiguous country. The whole region is too steep for cultivation, and habitations are few and far between and most dilapidated in character. The soil is stiff, clayey loam and the few crops we saw were wheat, Rye (*Secale fragile*), Irish potato, maize, and pulse. The cliffs are for the most part well timbered, and the common trees and shrubs of Hupeh are represented. *Pinus Armandi* is very abundant and *P. Henryi* is also common. Odd trees of Spruce and Hemlock also occur. A fine specimen of *Acer griseum*, 60 feet tall, 7 feet in girth, with curious cinnamon-red papery bark was the feature of the day's march ; unfortunately, it was badly situated for photographing. Beech, Yellow-wood and *Dipteronia sinensis* were common trees *en route*.

The road is one long succession of ascents and descents and most fatiguing. In the afternoon, after a particularly trying ascent, we wandered for an hour or so through woods of Oak (chiefly *Quercus variabilis* and *Q. aliena*) and Sweet Chestnut, the latter laden with its white, evil-smelling flowers. Walnut and Varnish trees are everywhere abundant and *Campanula punctata* is a common weed of cultivation. No foreigner had ever before traversed this region and the people were very timid, locking up their houses and hiding themselves from view at our approach. The cliffs in this neighbourhood are full of caves and many of these are bricked up to form places of refuge in troublous times. We found lodgings for the night at Peh-kuo-yüen, alt. 3750 feet, in the house of the head-man of the hamlet. Food-stuffs were scarce and there was great difficulty in persuading the people to part. What little we did eventually obtain was at famine prices and the grumbling was loud on all sides.

The following morning we descended by a moderately easy path to a torrent and then commenced a heart-breaking ascent of some 2600 feet. It was excessively hot and I do not remember perspiring so much before. A rugged, precipitous, sparsely populated country is this, and I never wish to see it again. Limestone regions are magnificent from the scenic point of view, but for travelling over they are fierce and arduous beyond words! Our destination was Hsao-pingtsze and no one knew the distance. Inquiries made as often as possible always elicited the same reply : " Seven or 8 li from Peh-kuo-yüen, 7 or 8 li to Hsao-pingtsze." Late in the afternoon the distance to go increased to 30 li and did not shorten until we suddenly sighted the two huts which form the hamlet of Hsao-pingtsze !

The ascent was largely under cultivation, but the final stage was through jungle. *Lonicera tragophylla* is common and was in full flower, but we saw no good plants. A bush of *Schizophragma integrifolium*, one mass of the purest white, on the cliffs, was conspicuous from afar. But the flora generally is very ordinary, with *Rhododendron discolor* and *R. Mariesii* common here and there. On reaching the top of the cliffs we entered a cultivated slope where Walnut and Varnish trees abound. The district is called Ta-ping-shan and consists of several scattered farmhouses surrounded by fields of maize, pulse, barley, and Irish potato. At one of these farmhouses my followers managed to secure a good meal and high spirits prevailed in consequence.

On leaving this place we continued to ascend by an easy path skirting rolling downs. A few scattered houses occur for a couple of miles but were mostly deserted, and we soon left all signs of cultivation and habitation behind us. The downs are treeless and clad mostly with grass with scattered bushes of Willow, Barberry and Spiræa. The depressions between the hills were masses of blue Forget-me-not. The whole region would make excellent grazing ground for cattle. Crossing over at 7950 feet altitude, we descended by an easy road for a mile or so and passed a couple of huts surrounded

by extensive plantations of Medicinal Rhubarb. Many fine
herbs luxuriate hereabouts, and among them *Iris Wilsonii*
with its yellow flowers was conspicuous, covering large areas.
Eventually we reached the edge of a precipice, down which
the road fairly tumbles for 5 li to Hsao-pingtsze. This
hamlet, as the name indicates, is situated on a tiny flat
(probably caused by a landslide) and boasts two miserable,
dilapidated houses. We took up lodgings in the smaller and
presumably less squalid of the two, but there were little to
choose between them in all conscience. On three sides the
hamlet is walled in by steep cliffs and the fourth is the edge
of a precipice itself. It was only some 30 yards from our hut
to the edge of this precipice, and the view from this point is
one of the most extraordinary and wonderful my eyes have
ever beheld. Below me (some 4000 feet the morrow proved)
at an acute angle lay a small village with a considerable river
flowing past it. Beyond this was range upon range of bare,
treeless, sharp-edged ridges, averaging 5000 to 6000 feet in
height, with outstanding higher peaks and grander ranges in
the beyond. The rocks are mainly of limestone, white, grey
and reddish, giving a bizarre appearance to the whole scene.
Never have I looked upon a wilder, more savage and less
inviting region. A storm was brewing and the light rapidly
failing, making it impossible to take a photograph, though
no photograph could have produced a picture that would give
an adequate idea of the savage grandeur of the whole scene.
It was indeed sufficient to awe and terrorize one. Such scenes
sink deep into the memory and the impressive stillness
produces an effect which is felt for long years afterwards.
Soon the angry rain-clouds darkened and blotted out the
whole scene and the next moment a thunderstorm burst over
us. This storm lasted through the night and, the roof of our
hovel being like a sieve, the rain soon converted the mud floor
of the hut into a quagmire. We huddled together and did
what we could to keep dry and warm, but the night proved
long and cheerless.

Soon after daybreak next morning we made our escape
from these wretched quarters, but rain was still falling, and of
the wonderful scene of the preceding evening nothing was

THE MARKET VILLAGE OF TAN-CHIA-TIEN

visible from the gap but an ocean of clouds. The descent is most precipitous and for the first 2000 feet we fairly tumbled down. Afterwards it became more gradual and led over a steep cultivated slope of red clayey soil, making walking difficult. Nowhere is this descent easy, and very glad were we all that our route was down instead of up this mountain-side. At the foot of the descent the road leads through a rocky defile to emerge on the banks of a clear-water river some 60 yards broad. Across this we were ferried to Tan-chia-tien, the village we saw from near our lodgings last night. This village consists of some fifty houses which are huddled together and overhang the river in front and cling to the cliff behind in an extraordinary manner. From this village a kind of long street with houses scattered here and there along its length extends for 2 miles to the village of Chikou, situated at the junction of this river with another of almost equal size. A mile or so from Chikou up the secondary stream are the salt wells of Taning-ching.

The road we struck at Tan-chia-tien is a highway leading northwards to Shensi and southwards to Kuichou Fu on the Yangtsze River. Hereabouts and down to Taning Hsien, 12 miles distant, and northwards I know not how far, the cliffs are sheer to the water's edge. The road is well graded and a good 6 feet broad, and has been excavated or blasted from the solid rock.

From Chikou to Taning Hsien is said to be 30 li with not a house or hovel between. To cover this we with difficulty engaged boats, long, narrow, lightly built affairs (Sin-po-tzu), turned up at prow and stern, with no oars and steered by long sweeps projected fore and aft. The current was strong and rapids numerous ; aided by a freshet we covered the whole distance in half an hour. The brief journey was through one grand chasm, the walls of rock being sheer to the water's edge with no space even for a shingle-bank to lodge. These cliffs are treeless and mostly bare with here and there grassy patches and clumps of delicate, graceful Bamboo (*Arundinaria nitida*). The road zigzags around the cliffs on the right bank well above high-water mark, and every inch of it has been blasted from the hard wall of rock. Stone gates and barriers occur at intervals, but there are no houses. This road is of such a nature that time and neglect can affect it but little, but it is now scarcely used except by occasional pedestrians and salt-carriers when the river is impracticable. I tried hard to discover when and by whom the road was built but found no one who could tell me. It is evidently one of the ancient arteries of China, and probably dates back to the discovery of the salt-wells. It struck me as being an old military road and may probably have been built centuries ago when Kuichou Fu was a place of infinitely greater importance than it is to-day.

The river I have mentioned, known locally as the Taning Ho, rises near the borders of Shensi, Hupeh, and Szechuan, and after flowing nearly due south enters the Yangtsze at Wushan Hsien. From Chikou boats descend to its mouth, 200 li distant.

Taning Hsien, alt. 750 feet, the most easterly inland town in Szechuan, is situated on the right bank of the river, here about 100 yards broad, and sweeping from the gorge in a fine curve. The town is wedged in on the side of a mountain-slope up which the city wall ascends for several hundred feet. The river-front is bounded on one side by the city wall, and the shops, houses and yâmens are crowded together near the river. The upper slopes enclosed within the city wall are given over to agriculture. The town, comprising about 400 houses, is the residence of a district magistrate, and boasts a trade in salt and odds and ends. Formerly it was the centre of a large opium traffic.

At Taning Hsien the Chinese Banyan (*Ficus infectoria*), so abundant and characteristic of the central parts of Szechuan, puts in an appearance. Near a temple, a few hundred yards from the north gate of the town, I observed from the boat what appeared to be a Mantzu cave built in the face of lime-stone rock. On inquiring I was told of four or five similar caves in this neighbourhood. Later I may have something to say about these caves, but it is interesting to be able to register their presence at the extreme eastern edge of the province, since heretofore they have been considered a feature of the more western parts. Physically and geologically speaking, the country east of the Taning River belongs to western Hupeh. Almost immediately west of it the characteristic red sandstone of Szechuan commences.

For twenty-two consecutive days my followers and I had struggled through the wild, lonely mountain fastnesses of northwestern Hupeh, suffering much from bad roads, worse accommodation and scarcity of food supplies. For the first time on record the journey had been accomplished by a foreigner, and one and all of my followers were happy in the thought of the comparative luxury and plenty of the country which was now before us.

CHAPTER VI

THE RED BASIN OF SZECHUAN

GEOLOGY, MINERAL, AND AGRICULTURAL WEALTH

THROUGHOUT the eastern and central parts of the province of Szechuan, from near the Hupeh boundary to the valley of the Min River, the predominant rocks are red clayey sandstones, probably of Jurassic age. These rocks are of immense thickness and impart a characteristic red colour to the surface, and for this reason the late Baron Richthofen gave the term " Red Basin " to the whole region. This basin is nearly triangular in shape, the city of Kuichou Fu marking the " apex." Imaginary lines connecting Kuichou Fu with Lungan Fu in the north-west, and Kuichou Fu with Pingshan Hsien keeping a little to the south of the Yangtsze River, respectively mark the northern and southern "sides." Another line from Lungan Fu and thence skirting the valley of the Min River to Pingshan Hsien marks the " base " of the triangle. The entire basin is nearly 100,000 square miles in area, and is surrounded on all sides by lofty mountain ranges, those on the west rising above the snow-line. In the east the boundary ranges are composed principally of Upper Carboniferous limestone, as described in Chapter II. The western boundary ranges are largely made up of shales. The Yangtsze River traverses the basin from west to east, following a course nearly parallel with the southern limits of the basin itself. Within this triangle there is abundant life, industry, prosperity, wealth, and intercommunication by water. Outside of it on all sides the contiguous country is sparsely inhabited, little productive and no river is navigable save the Yangtsze, where it leaves the basin.

In ancient geological times this region was doubtless a vast

inland lake with a fairly even floor.   Since the draining off of
the waters the Yangtsze River and its network of tributary
streams have eroded channels 1500 to 2500 feet deep through
these soft sedimentary rocks, and converted the whole basin
into a thoroughly hilly country.   To-day practically the only level
area is the Plain of Chengtu, some 80 miles long and 65 miles
wide, with an average altitude of 1800 to 2000 feet above sea-
level.   The rest of the basin is broken up into a network of
low, rolling or flat-topped mountains averaging about 3000
feet above sea-level, and nowhere exceeding 4000 feet altitude.
The whole of this region is under agriculture, the highest
development of which obtains on the Chengtu Plain, perhaps
the richest area in the whole of China.   Anent this particular
part we shall have something to say later.

How great a period of time has elapsed since the disappearance of the waters from this basin is purely conjectural.   But
that this triangle has long constituted a well-marked boundary
is evidenced by the fact that remarkably few of the plants
found in the mountains bordering the eastern limits at 2000
feet altitude and upwards are common to the mountains
bordering the western limits.   The genera are of course the
same, but the species are usually distinct.   The difference
between the floras of the eastern and western border-ranges is
too great for a mere 500 miles of longitude to account for
solely.   The same is true of the fauna in so far as the game
birds and animals are concerned, as Chapters XI and XIII,
Vol. II, dealing with these will confirm.

From evidence presented by the flora to-day it appears
doubtful if ever the Red Basin was covered with great forests.
Rather would I suppose that subsequent to the disappearance
of the waters the region bore some resemblance to the " bad
lands " of certain parts of the United States of America.
All this is admittedly pure conjecture.   Everywhere to-day,
trees, shrubs, and herbs are common, but the flora, in contra-
distinction to that of the contiguous regions, is relatively poor,
and the species largely common to the entire basin.   Further,
the majority of these species are widely spread throughout
the warmer low-level legions in China, some indeed ranging
to the extreme eastern limits of the country.   A theory is apt

to become fascinating, and may easily be carried too far. The facts above recorded are best left until the geology of China generally is more accurately known.

Coming down to historical times we learn that the region previous to the advent of the Chinese was peopled by an aboriginal population divided into the kingdom of Pa in the east and the kingdom of Shu in the west. This aboriginal population has entirely disappeared, but records in the shape of well-constructed caves having square entrances are found scattered all over the Red Basin. These caves are especially abundant around Kiating Fu. A little investigation of these interesting places has been attempted, and fragments of pottery and odds and ends discovered. The entrances to these caves could only be closed from the outside, and from this fact, and other details, it is probable that they served as the burial-places of the chiefs and more wealthy among this extinct people, rather than as dwelling-places or harbours of refuge. Doubtless they have been subsequently used for these latter purposes, but that they were designed for tombs seems to best explain their origin. From Chinese history we learn that as early as 600 B.C. the kingdom of Pa had relations with the Chinese kingdoms of Ts'u, which occupied the regions north of the barrier ranges. Later, Pa princesses married Ts'u kings. Ts'u was in time conquered by Ts'in (another Chinese kingdom), which gradually absorbed Pa, and finally conquered Shu about 315 B.C. A military road was commenced from the neighbourhood of modern Hanchung Fu, designed to connect with the region around modern Chengtu, by Ts'in-shih Hwang about 220 B.C. This road, which enters Szechuan from across the barrier ranges near Kuangyuan Hsien, is still in existence as the great highway connecting Chengtu with Hanchung Fu, Sian Fu, and, ultimately, Peking itself. For the next fifteen centuries the history of this region is full of war, rebellion, and internecine strife. Usurpers established petty dominion over the country from time to time, only to disappear amongst awful slaughter and bloodshed. There is scarcely a square mile of the whole region but what recalls scenes of valour, treachery, and carnage. In the latter half of the thirteenth century the famous Tartar, Kublai Khan, carried his arms victoriously

A TYPICAL VIEW IN THE RED BASIN

over nearly the whole of modern China, and formed an Empire which the succeeding Ming and Manchu dynasties maintained more or less intact.

Since the time of Kublai Khan many rebellions have swept over Szechuan, decimating the population and paralysing industry. The present population is mainly derived from immigrants (voluntary and otherwise) who settled there during the early half of the eighteenth century. A census taken in A.D. 1710 returned only 144,154 souls for the whole province. To-day the population is estimated at 45,000,000! In spite of all the long-sustained wars and bloody rebellions, agriculture has managed to subsist, and the whole of the Red Basin is a lasting monument to Chinese genius and industry in matters agricultural. An abundant water-supply and constant tillage are necessary to obtain a full crop from these sandy clays and marls. Fortunately, the whole region is one vast network of streams, all of which drain into the mighty Yangtsze. The Chinese have taken full advantage of this intricate river system, and devised manifold methods of irrigation. These devices, combined with the untiring patient industry of the people, have converted an incipient " bad land " into a rich and fertile region of terraced fields. In no part of China that I have visited are the people entitled to greater praise for meritorious agricultural accomplishment than throughout this Red Basin.

In many parts of this region the river valleys are so steeply eroded that very little cultivable bottom-land is formed. Consequently the rice belt is relegated to slopes and summits of the low, flattened hills. In limestone regions the bottom-lands constitute the main rice belt, but in the sandstone regions the opposite obtains. The climate of the whole region is mild and genial, and during both winter and summer the land is cropped. Rice is the great summer crop with maize, millet, sweet potato, sugar-cane, tobacco, pulse, and various other crops. The principal winter crops are wheat, rape, peas, broad beans, cabbage, Irish potato, etc. Formerly opium was cultivated in enormous quantities as a winter-crop, but this has lately been almost entirely suppressed. Cotton does not thrive in the Red Basin, though its culture is attempted in

many districts, notably Yilung Hsien and in Tungchuan Fu.
Cotton is the one commodity that this region has to import,
and nearly all its surplus products go to meet this deficiency.
But, if cotton is very little grown, many kinds of hemp are
produced in quantity, though very little is used for textile
purposes. Silk production is everywhere an industry of importance, and in many districts the staple. Only the very poorest
are without some silk garment, though such is only habitually
worn by the more wealthy. Tea is grown in many districts
both for local consumption and for export. In the more
westerly parts tea for the Thibetan market is a staple product.
Wood Oil and many other valuable economic trees are also
largely cultivated. Fruit is generally grown, including peaches,
apricots, plums, apples, pears, and oranges in variety.
Oranges thrive remarkably well in this red sandstone, and the
extensive orchards are a wonderful sight during the month of
December. Tangerine varieties are most generally cultivated,
and the fruit in season can be purchased at the rate of twelve
hundred or more for two shillings ! The tight-skinned varieties
are less frequently grown, and are more expensive. Around
Lu Chou are plantations of Litchi trees. When they came from
their original homes the settlers evidently brought with them
their favourite trees and grains and planted them around their
new homesteads. These introductions, and the favourable
climate, explain the presence of such a vast variety of cultivated plants, which is probably greater than that found in any
other province in China.

The steeper and rougher country is covered with small woods
of Oak, Pine, and Cypress, elsewhere trees are confined to the
vicinity of streams, houses, temple-grounds, wayside shrines,
and tombs.

The streams are navigable for extreme distances, and a
perfect network of roads traverse the basin in every direction.
These roads are, on the whole, well built for Chinese roads, but
are not kept in thorough repair any more than those elsewhere
in the land. The streams, however, are well supplied with
ferries, and well-built bridges, substantially constructed of
stone, and kept in good repair, are a feature throughout the
entire region. Large cities, market villages, hamlets, and farm-

A SOAP-TREE (SAPINDUS MUKOROSSI) 80 FT. TALL, GIRTH 12 FT.

houses dot the land, which everywhere appears prosperous and its inhabitants contented. Drought occasionally brings famine, but, on the whole, the Red Basin suffers much less from this dread calamity than do other and less favoured parts of the eighteen provinces of China.

The mineral wealth of the Red Basin is not varied, but enormous brine deposits occur scattered over the whole area, and are worked at depths varying from almost surface level to 3000 feet. In the eastern parts, Kuichou Fu, Wên-tang-ching, for example, the rivers have scoured the rocks until the brine-deposits are practically exposed. In the west, however, as at Wu-ting-chiao, situated on the left bank of the Min River a few miles below Kiating Fu, the brine is found at about 500 feet down. At Tzu-liu-ching, on the left bank of the To River, where the richest deposits occur, the brine is found at depths from 1000 to 3000 feet.

Salt is worked in some thirty-nine districts in the Red Basin. It is everywhere a Government monopoly, and its production and subsequent distribution are rigorously controlled. The annual output is estimated at about 300,000 tons. At Tzu-liu-ching most of the brine is evaporated by inflammable gas; in all other places the brine is evaporated by coal heat. In boring the deep wells, it is uncertain whether brine or gas will be struck, but both are equally valuable. The occurrence of this in-flammable gas indicates the presence of petroleum beds at still greater depth.

Coal is found in greater or lesser quantities scattered all over the Red Basin, and is always found not very far removed from brine pits. This coal varies from lignite to anthracite. The average quality is poor, but one or two good seams have been found, notably at Lung-wang-tung, a few miles north of Chungking.

Our early description of the Red Basin needs some ampli-fication to explain the presence of coal and other minerals. Although the sedimentary sandstones are in a state of undis-turbed stratification over a great part of this area, yet there is dissecting this Red Basin a number of linear elevations, in which the underlying limestone is bent up from a great depth. This limestone forms in every case an axial core, lined on either side by highly inclined strata, among which there is ordinarily noticeable, next to the axis, a double belt of coal-formation, followed on either side by strata of red sandstone standing on edge. Baron Richthofen estimates that " the area of the coal-bearing ground in Szechuan probably exceeds in size the total area of every other province of China." But probably throughout nine-tenths of this area the coal-measures are buried deep beneath the superincumbent strata, and with trifling exceptions can never become available for mining. In the linear elevations, above referred to, the belts of coal-formations, though narrow, are of great length. They are most readily accessible in those places where rivers have cut through and exposed the ends of the seams. Mining is done by means of horizontal adits working from an exposed surface inwards. Coal is very generally obtainable throughout the Red Basin, and is the ordinary fuel of the entire region.

Iron-ores occur scattered throughout the entire region, but though in the aggregate the iron-smelting industry is a considerable one, in no one place is iron made on a large scale.

Sulphate of iron (copperas) is found in combination with coal in one or two districts, notably in Kiangan Hsien. Lime is common to all the linear elevations mentioned above, occurring in juxtaposition with coal, and is burnt in kilns in the usual way.

Gypsum is found and worked in one or two places, notably Mei Chou and Pengshan Hsien, both districts on the Min River, between Kiating and Chengtu.

Mineral oil in small quantity occurs in the district of Pengch'i Hsien, where a native company has made some attempts to develop the industry, but with unsatisfactory results.

Other less important minerals occur in small quantities. The precious metals, gold and silver, are not found in the Red Basin proper but in the mountainous country to the west of this region, where copper, lead, and zinc ores also occur.

In reference to gold it should, however, be mentioned that rude placer mining is carried on during the winter months, throughout the numerous shingle banks exposed in the beds of the Yangtsze, Kialing, and Min Rivers. On the Yangtsze this precarious industry is first to be noted some 50 miles below

Ichang, but it is not general until the region west of the Gorges is reached. The industry is carried on by the unemployed peasantry, and the returns are most insignificant. This gold is in all probability brought down by the Yangtsze and its larger tributaries during the summer floods. There is no record of any gold-bearing quartz having been found *in situ* in the Red Basin proper. In the mountains bordering its western and north-western limits, gold quartz is found in greater or less quantities, and all the principal rivers of this region either take their rise in, or flow through, these ranges. This fact explains the presence of small quantities of gold far removed from the gold-bearing strata.

CHAPTER VII

EASTERN SZECHUAN

Narrative of a Journey from Taning Hsien to Tunghsiang Hsien

THE region described in this chapter was traversed by Lieut.-Colonel C. C. Manifold and Captain E. W. S. Mahon when surveying for a possible route for the proposed Hankow-Szechuan Railway in 1903 or 1904, I am not sure which. There is no record of any other traveller having crossed this part of eastern Szechuan, though it is very possible that missionaries may have done so. I do not know the conclusions arrived at by these surveyors, but the construction of a railway along the route I traversed would be a difficult and costly undertaking.

The following narrative is compiled from my diary, and may, perhaps, convey a brief idea of the nature of the country and the flora found in the more easterly parts of the Red Basin. As will be gathered, I took ten days to cover the distance, but I travelled leisurely, and the journey could be accomplished in six days.

June 28.—Yesterday we spent the day at Taning Hsien, refitting and preparing for our journey westwards to Chengtu Fu. Money exchange proved an involved and difficult business. Ten-cash pieces, both Hupeh and Szechuan currency, are accepted here at 20 per cent. discount. This means that the purchasing power of a thousand such cash is only equal to 800 string-cash. Farther west, Hupeh 10-cash pieces are not current, and the Szechuan 10-cash piece is only accepted for two days' journey west of this town. We had therefore to burden ourselves with string-cash, which added considerably to the weight of our loads. A thousand cash in 10-cash pieces weighs less than 2 lb.; in string-cash the same equivalent weighs over 8 lb.! If there is one reform more badly needed than another in China it most certainly is in the matter of currency.

Leaving Taning Hsien by way of the west gate we made a slight ascent and entered a narrow, highly cultivated valley, flanked on our right by fairly high and on the left by lower mountains, nearly treeless and sparsely cultivated. The town of Taning lies in a depression, and the morning mists obscured the general view. It is a very small place, with much of the land enclosed within its walls given over to cultivation. An outer gate, wall, and block-house guards the west gate proper.

Ascending the valley by an easy road which more or less skirts a fairly large tributary stream of the Taning Ho, we reached the village of Che-tou-pa before noon. Rice was abundantly cultivated everywhere, irrigation being effected by means of large " Persian " wheels. Much cotton is cultivated following wheat, the winter crop. Maize was 5 feet tall and in full flower. *Paliurus orientalis*, a thin tree 30 to 50 feet tall, is very common, and was laden with white, circular, odd-looking fruits. Weeping Willows, Cypress, and fine specimens of a hairy-leaved, small-fruited Hog Plum (*Spondias*) were noteworthy, with Bamboo groves in abundance.

On leaving Che-tou-pa we deserted the main tributary stream and ascended a small branch. The valley narrows, and the hills are more wooded, chiefly with Cypress. The road is easy, though here and there sadly in need of repair. We journeyed slowly, and eventually crossed over a ridge of low hills to the hamlet of Lao-shih-che, which we reached at 5 p.m. This tiny place, alt. 1950 feet, and 55 li from Taning Hsien, consists of half a dozen houses, scattered through a narrow valley with rice fields on all sides. The people were very nice, but inquisitive.

We were on the edge of the Red Basin and much of the soil had the characteristic red colour. Wood Oil trees are commonly cultivated, but cotton was not in evidence during the afternoon. In a grove I noted some magnificent trees of *Pistacia chinensis* and *Sapindus mukorossi*. The young shoots of the former are cooked and eaten, but the round fruits of the Sapindus are used as soap. Celtis trees are common, their smooth, pale-grey bark rendering them conspicuous. On a ridge we noted many trees of the interesting " Button tree " (*Adina racemosa*). These trees were 30 to 60 feet tall, 2 to 4 feet in girth, and the finest specimens of their kind I have met with. The Chinese Pine (*Pinus Massoniana*) is general, but by far the commonest tree of the day was the Cypress (*Cupressus funebris*).

The road proved a pleasant change ; instead of wild and savage scenery, low rounded hills backed by steeper mountains, all rather treeless, and for the most part cultivated, were the order of the day. Here and there were a few outstanding cliffs of limestone with an occasional temple crowning odd crags. At Taning Hsien we secured a number of new coolies, and these men described the country passed through in the afternoon as Laolin (wilderness). This immensely amused my Ichang men, who recommended these newcomers to try the Sheng-nêng-chia before speaking of " Laolin " !

The day was grilling hot, and all were fairly exhausted on arrival at Lao-shih-che. Whether it was the heat or the after effects of a day's holiday I could not determine, but I was called upon to play " Doctor " to nearly half my followers. The majority were suffering from stomach troubles, several from filthy sores. Epsom salts, permanganate of potash, and iodoform dressings soon improved the majority.

The next day was gloriously fine, but scarcely so hot as the previous day, or perhaps the slightly increased altitude made it more bearable. The whole day we travelled nearly due west through a narrow valley bounded by moderately high parallel ranges. The road continues easy with occasional ascents and descents. We were still on the fringe of the Red Basin, but in the afternoon grey sandy soils were most in evidence. Rice is cultivated wherever sufficient water is obtainable, and was scarcely ever out of our view. Maize is the other principal crop, with various kinds of pulse and the Irish potato. The sweet potato is cultivated here and there, and Wood Oil trees are even more abundant than before. Much oil is evidently produced in this region, and we noted many oil-presses during the day. The parallel ranges are

MAUSOLEUM WITH ORNATE MURAL SCULPTURING

from 500 to 1000 feet above the valley, sparsely cultivated, and for the most part well timbered with Cypress (*Cupressus funebris*), Pine (*Pinus Massoniana*), and Oak (*Quercus serrata*). Poplar is a common tree, and by the sides of streams Weeping Willows abound. Shrubs in variety occur the most noteworthy being *Itea ilicifolia* and *Torricellia angulata*. Nowhere else have I seen this latter shrub so plentiful; it favours the sides of streams, ditches, and rocky gullies, forming a densely leafy bush 8 to 12 feet tall. The fruit when ripe is black, and is borne in large pendulous cymes. The Itea occurs in rocky places, and its pendulous tails of greenish-white flowers are often 18 inches long. The leaves very closely resemble those of the common Holly, and when not in flower it might easily be mistaken for that plant.

Houses are scattered along the route, but the population is sparse. We met a few mule trains, but there was really very little traffic on the road. We found accommodation for the night at Hsia-kou, a prettily situated hamlet, alt. 2800 feet, 65 li from Lao-shih-che. Our lodgings were spacious, but the occupants of the house looked unprepossessing opium sots.

At To-chia-pa, a small hamlet passed a few miles before reaching Hsia-kou, a road branches off to the northward and leads to Chêngkou Ting. It was said to be a hard road to travel over.

On leaving Hsia-kou we immediately plunged into a ravine with steep limestone cliffs 300 to 500 feet high; the road follows the dry bed of a torrent. At the head of this ravine we made a slight ascent, and wandered across low mountain-tops for a few miles, then descended and crossed a branch of the Kuichou Fu River by a covered bridge. Up to this point Pine and the Chinese Fir (*Cunninghamia lanceolata*) are common. At the bridge I photographed the largest tree of *Platycarya strobilacea* I have seen. This specimen was fully 75 feet tall, with a girth of 6 feet. I had no idea it could attain such dimensions. A few miles beyond this point we forded the main branch of the Kuichou Fu River, a broad, shallow, clear-water stream, and about noon reached the village of Chiao-yang-tung. Soon afterwards we were overtaken by a furious thunderstorm, which arose with amazing suddenness. The fury of the storm spent itself in a torrential downpour of short duration, but rain fell steadily during the rest of the day. The rain did not improve the mud road, and our progress was slow and difficult in consequence. During the whole afternoon we made a steady ascent, skirting the mountain-sides through woods of Pine and Oak. Eventually the road enters a narrow sloping valley, at the head of which we found lodgings for the night in two houses which constitute the hamlet of Shan-chia-kou, having travelled 65 li. Around this place the flora is varied and essentially cool-temperate in character. Bushes of Mock Orange (*Philadelphus*) were conspicuous on all sides with their wealth of pure white flowers. The Hautboy strawberry is abundant, and around our hostel I gathered in a few minutes enough of these luscious fine-flavoured white berries to stew for dinner. The Torricellia was again common. It ascends up to 3500 feet altitude, and often forms a small inelegant tree.

We saw very little rice during the day, maize and Irish potato being the chief crops. There is practically no traffic on this road; the mule-trains seen yesterday evidently came down the road from Chêngkou Ting. Population is sparse, and what there is looked strongly addicted to the opium habit. So far, however, we had not seen any signs of poppy.

A magnificent day ushered in the new month. The morning was bearably hot, but the afternoon scorchingly so. A hundred yards beyond our lodgings we reached the head of a ridge, and an abrupt descent of a couple of thousand feet or so led to a narrow valley where much rice, maize, Irish potato, and a little Hemp (*Cannabis*) are cultivated. The parallel ranges flanking this valley are of limestone with outstanding bare rocks and cliffs, very little cultivated but with good woods of the common Pine. Here and there in the valley we passed fine trees of Sassafras, Sweet Chestnut, Sweet Gum (*Liquidambar*), Chinese Fir, and Poplar. At the head of the valley we made a slight ascent to the top of a ridge. Below us, some 2500 feet, flowed a considerable river walled in by lofty limestone precipices. It was 10.30 a.m. when we reached the top of this ridge, and the rest of the day's march was a more or less precipitous descent to the river, which we reached at Sha-to-tzu about 3 p.m. In its early stages the descent is as difficult as it well could be—over loose Rowley-raglike debris, down and up steep steps, and over slopes of greasy clay. We crossed one or two cultivated slopes, but most of the time the road skirts around the sides of cliffs. At the edge of one precipice, 500 to 1000 feet sheer, the road is carried through a narrow tunnel some 50 yards long and 3½ feet broad at the exit. This tunnel is partly natural and partly made by blasting the hard limestone. It was quite dark within the tunnel save for a faint glimmer of light at the exit. Both chairs and loads were with difficulty carried through this tunnel. This roadway is of recent date, and is unique in my experience of Chinese roads. Rough as it is it saves about 10 li and a very steep ascent and descent.

From the tunnel-way the road skirts the tops of the cliffs with many exasperating and wearying ascents and descents. Finally we descended to a small tributary of the main stream and, crossing over, reach Sha-to-tzu, a busy market village and, for the nature of the country, of considerable size. Up the tributary stream some 10 li, iron is mined and smelted, the quality being described as good. Around Sha-to-tzu, coal is worked and lime burnt.

The river we had with so much fatiguing travel reached enters the Yangtsze at Yunyang Hsien, distant 150 li. It is a clear-water stream of considerable volume, and is navigable for small boats from just below Sha-to-tzu to within 15 li of its mouth. Salt and a little peddling traffic was noticeable on the road; also odd loads of medicines, including Tu-chung, the bark of *Eucommia ulmoides*. The salt is a product of Yunyang Hsien, and is not allowed to enter Taning Hsien. The quality is said to be superior to that found within the latter district.

The flora of the day's journey was not particularly interesting, being very much the same as that found in the glens and gorges around Ichang. A new Stachyurus and *Abelia Engleriana* were collected. The Heavenly Bamboo (*Nandina domestica*) was particularly abundant in rocky places, its elegant foliage and large erect trusses of white flowers with conspicuous yellow anthers making it very attractive. In autumn and winter the masses of scarlet fruit render it extremely beautiful.   Wood Oil trees were general in rocky places, and *Hypericum chinense*, a wealth of rich golden yellow, was strikingly handsome, nestling on the cliffs everywhere.   Quite a little Ramie (*Bœhmeria nivea*) is cultivated, and the people were busy stripping the fibre-containing bark from the stems. The leaves, like those of several other plants, are used for feeding pigs.   The stripping and cleaning is all done by hand labour.

The day's march was full of interest, but the intense heat and hard road made the 60 li very trying, and all were glad when the end of the stage was reached.   The scenery was magnificent, and forcibly reminded us of the glens and ravines around Ichang.   The railway surveyors must have been filled with despair when they encountered this steep limestone country !

Sha-to-tzu is only about 700 feet altitude, and, in spite of the swift-flowing stream which passes its "front door," was suffocatingly hot.   We managed to find a good inn with quarters removed from the street and remarkably private in character.   We had no difficulty in changing silver here, but 10-cash pieces are no longer negotiable.   String-cash was the only kind the people would accept.

Just below Sha-to-tzu we crossed the river by a ferry which is assisted by a convenient rapid, and commenced a steep ascent.   A few hundred feet up we were afforded a good view of the village we had just left.   It contains about a hundred houses, crowded together on a narrow, fan-shaped slope.   A few temples shaded by large Banyan trees were conspicuous, and the whole made a decidedly pretty scene.   The ascent is through cultivated fields, groves of Wood Oil trees, and finally Pine woods.   At 3100 feet altitude we crossed a gap, and 200 feet more led to the top of the range.   The rest of the day we followed an undulating, easy road which meanders through rocky, Pine and Cypress clad mountain-tops, and finally descends to Chê-kou-tzu, which was our destination for the day.

The country is very pretty ; farmhouses are scattered along the route, and where possible the land is under cultivation. Rice was of course *the* crop where water is obtainable, maize

TANING HSIEN : NORTH GATE

and Irish potato elsewhere. Tobacco is grown; a little of this crop has been noted every day since leaving Taning Hsien. Limekilns were common all day. In one place we saw a number of men out with guns after Muntjac. They fired several times, but did not succeed in killing the animal during the time we watched the sport.

A few li before reaching Chê-kou-tzu we passed an unusually large house of much architectural beauty. It was erected by a rich man named T'ao, who held the purchased rank of " Hsien." He died some twenty years ago, and the family has fallen on evil times, thanks to idleness and opium.

The flora was not very interesting. Some fine trees of Cypress and odd ones of *Catalpa Duclouxii* were noteworthy. Pine abounds, and I saw several examples of " clustered cones." These cones, a hundred and more crowded together, were all small, and appeared to have displaced the male flowers. Chê-kou-tzu, alt. 2050 feet, consists of some forty houses situate above the mouth of a stony stream and backed by low mountains, on the top of which is an ancient fort.

On leaving Chê-kou-tzu we immediately entered a pretty valley, highly cultivated with rice and bounded by low, rolling hills. A large number of farmhouses and a small hamlet occur in this small but prosperous valley. Throughout the whole forenoon we traversed a number of such depressions separated one from another by low ridges, always ascending slightly with the valleys narrowing until finally they become mere basins surrounded by rocky limestone mountaintops. Crossing a final ridge we entered Kai Hsien at a place called Shih-ya-tzu, 35 li from Chê-kou-tzu. Up to this point the scenery is very pretty, the rocky mountain-tops being clothed with woods of Pine and Cypress. Oak is common, and in more open places and around habitations we passed fine trees of Spondias, Pistacia, Paulownia, Tapiscia, and *Hovenia dulcis* ; the last covered with masses of white flowers.

The afternoon's journey was all downhill, ending in a very precipitous descent to Wên-tang-ching. The road led through maize plats, odd rice fields, and bare, treeless hilltops, with no flora of interest. Nearing our halting-place for the day it was fearfully hot, and the absence of shade was severely felt.

Here and there the hilltops are crowned by old forts built of dressed stone. These relics (Chaitzu) of turbulent times abound all over the salt districts and more wealthy regions of eastern Szechuan. Limekilns, small clay-covered affairs, were common *en route*, and many of the rice fields had been dressed with slaked lime.

Wên-tang-ching is a town of considerable size, by far the largest place we had met with since leaving Ichang. It is built on steep slopes bounding the two sides of a clear-water stream, and backed by high limestone cliffs. On the south-west side these cliffs are stark and sun-baked. Large quantities of salt are produced here. The brine pits are situated on the foreshore and immediate neighbourhood of the stream. The supply depends on the state of the river, the lower the water the more brine is obtainable. During summer floods the industry is suspended. The salt is white, powdery, of moderate quality, valued at twenty-six cash per 16-ounce catty. It is distributed throughout the north and west of the Hsien, but cannot enter the city of Kai Hsien itself. Dust-coal is mined in the neighbourhood and used for evaporating the brine.

The town consists of about a thousand houses and boasts several temples and large guild-halls, that belonging to the Shensi guild being very prominent on account of its large size and ornate architecture. Two small pagodas protect the luck of the place, and many Chaitzu crown the surrounding hills. The inhabitants are not prepossessing, being unusually dirty and over-curious. Some were not over-civil, and there was a slight scuffle between my men and some rowdies. Our inn was dark, suffocatingly hot, and most undesirable in every way. It was the best we could find, and served its purpose, uncomfortable as it was. Behind the inn is a huge cave with vast stalactites and a cool breeze blowing through it. This is the curiosity of the town, and was pointed out with a great show of pride.

All along the route from Taning Hsien there has been much argument over the price of food-stuffs. The natives constantly putting up the price on my men, this led to heated words, but generally ended in the men getting a fair price. Many of them had travelled too far not to know "the ropes."

Wên-tang-ching is only 750 feet altitude, and with the

heat from stark surrounding cliffs and hundreds of furnaces is a regular inferno. Prosperous it may be, but it failed to appeal to us, and one and all were glad to quit it.

A steep ascent of a few hundred feet and we cleared the town. After passing through a large graveyard we descended to an alluvial valley where much sugar-cane, maize, tobacco, and a little cotton is cultivated. The road is broad, paved with blocks of hard stone, and traverses the valley to its head at Ma-chia-kou, 12 li from Wên-tang-ching. Ma-chia-kou is the coal port for the salt-wells. Coal is carried overland some 30 li, and at this point put into small boats and conveyed to the brine-pits. This coal is valued at three cash per catty, the carriers receiving one cash per catty for carrying it down. The boats are small, steered by sweeps fore and aft, and can descend this stream to Kai Hsien, 60 li below Wên-tang-ching, and from thence to Hsiao Ch'ang on the Yangtsze, 110 li distant. At Ma-chia-kou the road leaves the main stream, which flows down from the northward, and after crossing a neck descends to a broad stony torrent, which it ascends through uninteresting country, eventually leading through a limestone ravine. The coal supply is of primary importance to the salt-wells, consequently the road is kept in good repair. During the forenoon we met hundreds of coolies and many women laden with coal. Iron is found in this neighbourhood, and pigs of this metal were being carried down to the boats.

On leaving the above-mentioned ravine we traversed a valley of rice fields and reached Yi-chiao-tsao about noon. Five li above this hamlet we crossed over, and during the rest of the day's march descended a narrow valley flanked by steep Cypress-clad slopes. Sweet potato is abundantly cultivated, also rice and maize. Houses are frequent, and the people appear fairly well-to-do.

We found lodgings for the night at Wang-tung-tsao, alt. 1350 feet, having covered our usual 60 li. The day was terribly hot, making the journey very fatiguing. The inn is beautifully situated in a grove of Bamboo and Cypress, but is poor and abominably stinking. Really, it is a pity that such a vile house should defile such a charming spot.

The next day was also grilling hot, with no signs of a storm to cool the air.   Descending a few li we struck a rather broad stream with many red-sandstone boulders in its bed.   The road ascends this stream to its source, and steep ascents and descents were all too frequent.   We lunched at the village of Kao-chiao, and a more hot, fly-infested, stinking hole, with people more inquisitive, I have not experienced.   Savage, snarling, yelping dogs abounded, and these, with the other discomforts, did not add relish to the meal.   My followers seemed to share my views of this village, and grumbling and malediction were loud on all sides.   Our meal did not occupy long, and we all felt better when clear of this filthy, pestiferous place.

The whole day was spent among sandstone, grey and red, and we were seldom out of sight of rice fields.   Pine abounds, but the Cypress does not appear to be at home here, and occurs very sparingly as compared with previous days.   Wood Oil trees are common, but the flora generally is not interesting.   Elæagnus bushes are common and were in ripe fruit.   The stems of this shrub (Shan-yeh-wangtzu, or Yang-ming-nitzu) are commonly used for making the long stems of tobacco pipes so frequently seen in this region.   The Burdock (*Arctium major*) is common in stony places and often cultivated, being used as medicine under the name of " Yu-pangtzu."

Three li before reaching our lodgings we crossed a ridge, and passing through a stone gateway, entered the district of Tunghsiang Hsien.   We found an inn at P'ao-tsze, a small scattered hamlet, alt. 2650 feet, 65 li from Wang-tung-tsao. The inn is clean and prettily situated in a little valley bounded by low red-stone hills all under cultivation.   The host is evidently a man of substance, and amongst other things owns a reclining chair of novel workmanship, of which he is evidently very proud.

There was no breeze last night, and I slept badly, partly owing to the heat and partly to the occupants of the inn talking in high argumentative tones till past midnight.   This is a common habit of the Chinese and very exasperating to any one trying to get to sleep.

With only 50 li to do to Nan-pa ch'ang the men were in high spirits and set out in style.   The road proved easy—by one o'clock we had covered the distance, and had a couple of

THE CHINESE PISTACHIO (PISTACIA CHINENSIS) 60 FT. TALL, GIRTH 25 FT.

long rests into the bargain. On leaving P'ao-tsze we made a short, steep ascent, and then descended by an easy road leading over and among sandstone bluffs. Twenty-five li on we reached the bed of a small stream and followed it to its union with a large, clear-water stream flowing down from the northward. This stream flows past Nan-pa ch'ang and is navigable for small boats down to Tunghsiang Hsien and up-stream some 290 li to Tu-li-kou. Near our destination we passed many coolies carrying down bright anthracite coal. This comes from Fu-che-kou, some 50 li away, and the men receive 200 cash per picul (100 catties) for carrying it down. We also noted iron in flat slabs, which comes from Tung-che-kou, 25 li distant.

Pine was again the common tree, but Cypress also was fairly common. The sandstone is evidently more favourable to the Pine than to the Cypress. We saw two or three trees of the rare " Hung-tou-shu " (*Ormosia Hosiei*). The wood of this tree is highly valued and so heavy that it sinks in water. Wood Oil trees continued abundant, and around Nan-pa ch'ang plantations of Mulberry were being made. Evidently sericulture is about to be attempted in this district.

Nan-pa ch'ang, alt. 1550 feet, is a village of considerable size, and is built on a flat bordering the stream. Formerly it was one of the most important centres of the opium trade in Szechuan, and its product was of very superior quality. The opium trade is now completely stopped, and this place has suffered tremendously in consequence. It also boasts a trade in general merchandise, supplying a large area of country to the northward. But opium was its real source of wealth, and with the disappearance of the opium traffic all trade has declined. To the northward a lot of tea is grown and the leading people of Nan-pa ch'ang are endeavouring to divert this trade from its present headquarters, Taiping Hsien, to their own village.

Around Nan-pa ch'ang there are a few Mantzu caves. Everything was very quiet in the village and we attracted little or no attention. We saw a couple of uniformed police, odd street lamps, and other signs of modern ideas. Leaving this village the next morning at 7 a.m., in four small boats, we dropped down the beautiful clear-water stream, and reached

Tunghsiang Hsien at 3 o'clock.   The distance is 140 li by water, 90 li by land.   Numerous rapids obstruct the stream, but since the volume of water is comparatively small they are not dangerous.   The river is bounded by sandstone cliffs, often steep and covered with Pine, Cypress, and mixed shrubby vegetation.   Chaitzu are common, and here and there we passed villages.   Cultivation is general, and the crops were beginning to show signs of suffering from drought.   Pulse in variety is abundantly cultivated, together with rice and other favourite articles of food.   Ordinarily the whole region is one of plenty and prosperity.

It was a pleasant change dropping swiftly down this beautiful river, and we all enjoyed the journey.   On reaching Tunghsiang Hsien a thunderstorm broke and the rain cooled the air delightfully.

We entered the city of Tunghsiang, alt. 1400 feet, through the east gate, and found accommodation in a quiet and moderately clean inn.   The city, though not large, seemed a busy place.   Formerly it boasted a large traffic in opium, and its general trade was then very considerable.   It nestles among low hills on the right bank of the river, and is faced on the opposite bank by steeper and higher mountains.   Sandstone cliffs and bluffs abound, and in some respects the whole scene reminded me of the country around Kiating Fu.

Our inquiries into the matter of currency disclosed the fact that Szechuan dollars are accepted here, but 10-cash pieces were still useless.   The Roman Catholic and China Inland Mission have established outstations here.   An Irish missionary belonging to the latter was staying here at the time of my visit, and I enjoyed for an hour or so the pleasure of his company. It was pleasant to hear my own tongue spoken again.   Not since leaving Ichang, 35 days before, had I encountered a single foreigner.

CHAPTER VIII

THE ANCIENT KINGDOM OF PA

NARRATIVE OF A JOURNEY FROM TUNGHSIANG HSIEN TO PAONING FU

FROM Tunghsiang Hsien the recognized route to Chengtu or Paoning Fu descends the river via Suiting Fu to Ch'u Hsien, then strikes westward to Chengtu, north-west to Paoning Fu. I had no fancy for the main route, since, by going due west from Tunghsiang Hsien to Paoning Fu, we should explore new ground. My map (War Office, Province of Ssu-ch'uan, Eastern Sheet) gave no route, but indicated villages, and it was evident, therefore, that these villages were connected by a road of some sort. Chiangkou seemed a good place to start for, so my men were instructed to find a cross-country road to this town. At first the innkeepers, chair hongs, and local officials denied all knowledge of any such road, and indeed of such a place. But any one who has travelled in China values such denials at their proper worth and is not discouraged. The men who had charge of these inquiries were trusted followers of ten years' standing, and though entirely ignorant of the geography of the region could be relied upon to ferret out a route if such existed. After about six hours' investigation, from the magistrate's Yamên downwards, I was informed that a small mountain road did exist, but was over hard and difficult country, affording the poorest of accommodation. This was sufficient ; they were told to get an itinerary of this route and engage a few local men as extra carriers. I went to bed about 10.30 p.m., satisfied that by 6 o'clock next morning everything would be ready for our cross-country jaunt. In my travels about China I have been singularly fortunate in never having any trouble with the Chinese. In the spring of 1900

I engaged about a dozen peasants from near Ichang. These men remained with me and rendered faithful service during the whole of my peregrinations. After a few months' training they understood my habits thoroughly and never involved me in any trouble or difficulty. Once they grasped what was wanted they could be relied upon to do their part, thereby adding much to the pleasure and profit of my many journeys. When we finally parted in February 1911, it was with genuine regret on both sides. Faithful, intelligent, reliable, cheerful under adverse circumstances, and always willing to give their best, no men could have rendered better service.

This cross-country journey from Tunghsiang Hsien to Paoning Fu, via Chiangkou, promised to be of more than ordinary interest. There was a novelty about it also, since there was no record of any foreigner having attempted it before. The route lay across the old kingdom of Pa (see Chapter VI, p. 66), and I hoped to find some evidences of this ancient race. Chinese history is dry, difficult reading, and it is hard to dig out solid facts. Wars, rebellions, and massacres deluge everything in blood ; the arts of peace are seldom given any prominence. The Chinese historians have always treated the aboriginal races with arrogant contumely, rendering it almost impossible to discover at this late date anything about the arts and life of these lost peoples. That the modern province of Szechuan boasted kingdoms and dynasties of its own before the advent of the early Chinese is historical fact. The first Emperor of the Ts'in dynasty, Tsin-shih Hwang or Shih Hwang-ti (221–209 B.C.), incorporated part of the kingdom of Pa with the rest of his dominions and nominally also that of Shu, whose capital was near modern Chengtu Fu. The succeeding Han dynasty (206 B.C. to A.D. 25) made the conquest complete. Since this time no aboriginal chief has ruled the Red Basin of Szechuan, though it has been conquered and re-conquered time and again by usurping Chinese and alien races. During the period A.D. 221–265, the Chinese Empire was divided into three kingdoms, one of which, under the Emperor Liu-pei, had its capital at Chengtu. Liu-pei and three of his generals and statesmen are handed down as popular idols, and everywhere in Szechuan

A SANDSTONE BRIDGE WITH CYPRESS AND BAMBOOS

stories are told of the doughty deeds accomplished by these heroes of old.   With this brief introduction I again take up my narrative :—

My principal men once more proved equal to the occasion, and on 8th July everything was arranged for our cross-country journey.   An itinerary had been made out, and the Hsien provided us with a couple of uniformed soldiers.   (He sent six, but I managed to get them reduced to two.)   Heretofore on this journey we had managed to avoid taking official escort, although it is the custom to do so in Szechuan.   No ordinary traveller desires this honour, but it is thrust upon him and cannot easily be avoided.   The presence of this escort renders the officials responsible for the traveller's safety in accordance with treaty arrangements.   It is necessary to pay these men a few cash, but often they prove useful in odd ways.   Cash is cheap, and an extra hundred per day for each soldier does not amount to any considerable sum.   The difficulty is in keeping the escort down to two men.   Four and six are common numbers, and if one did not protest continuously an almost unlimited number of authorized and unauthorized ragamuffins would attach themselves to one's caravan.   If there is cash to be made the legitimate escort is often not above farming in a few extra " hands," thus securing more money.   The escort is provided with a letter from the official supplying it wherein the number of men dispatched and their destination is given, so by examining this it is possible to check any attempt at fraud. On dismissing these men at their journey's end it is necessary to give them a card to carry back to their superior.   Their letter is stamped by the official who provides the new escort, and the card signifies that their duty has been satisfactorily carried out. If they return without a card for any reason or other they are liable to be punished.

Leaving Tunghsiang Hsien by the west gate we followed the main road to Suiting Fu for a few li, then branched off to the right.   The road is well paved, and we met plenty of traffic. For the first 20 li the road is practically level, winding in and out among low hills.   It then makes a steep ascent to the top of some bluffs, where Mien-yueh ch'ang is situated, 30 li from the city of Tunghsiang.   Throughout the rest of the day the road was easy, leading through and among low hilltops and shallow valleys intercepted by hills 300 to 500 feet high.   Cypress and Pine are abundant, so also are Pistacia and *Albizzia lebbek*, both making large umbrageous trees.   *Vitex Negundo* is the commonest shrub, sometimes attaining to the dignity of a small tree : it was everywhere covered with masses of lavender-purple flowers.

The country is highly cultivated.   Rice predominates, with various kinds of beans (especially Lutou, *i.e.* green beans) next in importance—both crops evidently follow after wheat.   We passed odd patches of cotton and very many Plum trees. The region is well populated, bypaths abound, and it was no easy matter for us to keep to the right road.   At one point the road bifurcates, one branch leading to Shuang-ho ch'ang, the other to Shuang-miao ch'ang, our proposed halting-place for the night.   The names of these two villages, when spoken rapidly, sound much alike, even to Chinese ears.   My men got somewhat confused, and for a time there was danger of the caravan following two divergent routes.

We passed through the market village of Wang-chia ch'ang (ch'ang signifies market village), a curious little place, dominated by a temple in the middle, the roofs of the houses uniting to form a central covered way, beneath which the road passes through the village.

Shuang-miao ch'ang was our intended destination for the day, but being market day the village was filled to overflowing. A hundred or more people followed us into an inn, and in a little while there was hardly room to breathe.   Many were obviously under the influence of wine.   It was too hot to tolerate such overcrowding curiosity, so we pushed on a further 5 li, where we happened on a decent farmhouse, which we commandeered.   The owner being away, his wife was at first sorely afraid, but in a couple of hours her confidence was gained and all was well.   The men had difficulty in obtaining food and lodging.   The majority went back to the village, but none complained : they all realized the impossibility of my remaining the night in such a crowded place.

Our quarters were new and shaded by a grove of Bamboo and Cypress, but mosquitoes were multitudinous, rendering life miserable. The place is called Hsin-chia-pa, alt. 1950 feet. We had covered 80 li, through a rich and interesting country. Lady Banks's rose was particularly abundant, with stems 2 feet round, festooning trees 40 to 50 feet tall. Mantzu caves occur sparingly. In several places we passed cultivated patches of *Panicum crus-galli*, var. *frumentaceum*.

We parted excellent friends with our hostess at Hsin-chia-pa, a trifling present and 400 cash made her extremely happy ; her thanks were both genuine and profuse. Soon after starting we made a precipitous ascent of 1000 feet and crossed what is probably the water-shed of the Suiting and Sanhuei Rivers. A descent led to the head-waters of a small river, where is situated the tiny market village of San-che-miao. Market was in full swing, the one short street with its few hovels being crowded with people. We passed through without stopping to satisfy the curiosity of the crowd. At this village several roads converge, the one we followed continuing to descend the stream, and leading through a rocky jungle-clad defile. The cliffs are of red and grey sandstone, steep, rugged, and crowned with Pine and Cypress. As fluviatile shrubs *Distylium chinense*, various Privets (*Ligustrum*) and *Cornus paucinervis* abound. The last-named is a low-growing shrub with spreading branches, and laden with small flat corymbs of white flowers it formed a most attractive bush by the water's edge. In the jungle-clad slopes through which the road winds Tea bushes 15 feet and more tall are common. They looked uncommonly like spontaneous specimens, but were possibly planted long ago, though some of them have been undoubtedly naturalized. Occasional trees of the Red Bean ("Hung-tou"), *Ormosia Hosiei*, occur ; at one time this was probably a very common tree in this region. Its timber is most valuable, and the tree has been ruthlessly felled. There is practically no cultivation in this defile, or room for any, and not a house for 20 li.

After traversing this wild and interesting ravine for several hours we made a steep ascent to the top of the cliffs, and on the way up discovered spontaneous plants of the Tea Rose (*Rosa indica*) in fruit. These were the first really wild specimens I had met with. Once on top of the cliffs we found that the country all around is under cultivation, chiefly rice, with houses at frequent intervals.   After a few li the road descends to the river again, and crossing by stone steps we reached the market village of Peh-pai-ho, where we found accommodation in a large house.   This village, alt. 1600 feet, also known as Peh-pai ch'ang, is a small place with unprepossessing residents. Our quarters were dark, fairly filthy, and loafers crowded around until bedtime.

The day's journey of 60 li was through a sparsely populated country, which, considering the low altitude, was unusually wild and jungle-clad.   The flora had points of interest, the finding of Tea bushes and bushes of the Tea Rose in the rocky defile being particularly noteworthy.   On bare sandstone cliffs large white trumpet-flowered Lilies were common, with their stems thrust out at nearly right angles to the cliffs. We met very few people on the road, and most of the women we saw had natural feet.   In the early morning we passed quite a lot of *Panicum crus-galli*, var. *frumentaceum*, cultivated.

The itinerary my men secured at Tunghsiang Hsien did not err on the side of accuracy.   Constant inquiries were necessary, but the results were confusing.   The river which flows past Peh-pai ch'ang was said to unite with the Chiangkou stream at Chiang-ling-che, 70 li distant.

A heavy thunderstorm occurred in the night, accompanied by a downpour of rain which lasted intermittently into the early forenoon of the next day.   The country needed rain badly, and the air was cool and fresh in the morning.   Peh-pai ch'ang is a regular warren of dilapidated houses, filthy and stinking, with a loafing and unduly curious population.   A loin-cloth belonging to one of my chair-bearers was stolen during the night, and my followers had little that was com- plimentary to say about the village or its inhabitants.

Following the river down-stream for 5 li, we reached Lei- kang-k'êng of the maps.   This hamlet (pronounced Lei-kang- t'an, from a fine waterfall on a small river which, flowing from the north, joins the main stream at this point) consists of a deserted temple, a few scattered houses, and an old fort high up on the cliffs.   It and Ta-chên-chai, another old fortress, are the only places marked on the map—both are

THE MARKET VILLAGE OF TAI-LU CH'ANG

to-day of no importance. The market villages, the real places of importance, are not shown. Maybe these villages have sprung up comparatively recently, and the forts, from long-continued peace, lost their importance. This is the only feasible explanation which occurs to me. This section of the country is only known from Chinese maps, and these were probably compiled during military times long ago.

From Lei-kang-k'êng a steady ascent for 30 li leads to the top of a ridge where is situated the important market village of Peh-shan. This place boasts a fine temple and about a hundred houses. Like all such villages in these parts it consists of one central street, practically closed over by the nearly uniting eaves of the houses. These market villages are a striking feature of this part of Szechuan. They are situated approximately 30 li apart, and nine markets are held monthly in each. These are arranged in such manner that the three villages lying nearest to one another hold market on different days, thus between them practically covering the month. On market days the country-folk assemble from all sides to buy and sell. Pedlars and itinerant merchants constantly journey from market village to market village. Such markets are of the highest importance in a sparsely populated country, but the denizens of these villages suffer from too much spare time. Market days are what they exist for, and on : the other days are mainly spent in gambling and sloth. This system of market villages dates away back to the very dawn of Chinese civilization, and in the region we are concerned with here, is very little changed from what it was in the earliest times.

Five li before reaching Peh-shan ch'ang we struck a road which comes from Suiting Fu, 120 li distant. The country hereabouts is split up in low mountain ranges, averaging 3000 feet altitude, composed of grey and red sandstones. The river-valleys are mere ravines clothed with dense jungle, Pines, and Cypress, with no bottom lands nor cultivation of any sort. Some 500 feet up the cultivated area begins and extends to the summit. Terraced rice fields abound, tier upon tier, intercepted by low bluffs, the tops of all of which are cultivated. The whole country is very pretty, and in many respects peculiar, as far as my experience goes. Most of the women have natural feet, and many were busy weeding and firming the rice plants.

On leaving Peh-shan ch'ang the road makes a steep descent to a stream and a correspondingly steep ascent to the top of the bluffs again, winding round to the crest of a ridge where is situated the market village of Yuen-fang. This place, alt. 3100 feet, which was our destination for the day, having covered the allotted 60 li, is prettily situated. We found lodgings in a new and clean house boasting a veranda overlooking a grove of Pine and Cypress trees. The crowd which collected was small and though inquisitive kept at a respectful distance.

The flora proved identical with that of the previous day's journey. I again met with sub-spontaneous Tea bushes in the jungle and also saw a number of the Red Bean trees. Perhaps the most interesting objects noted during the day were the tombstones. These are very different from any I have seen elsewhere. They are of freestone, often highly sculptured, the workmanship being superior and the effect both artistic and dignified. One or two old stone mausoleums were magnificently sculptured. The aboriginal population of this region were accomplished workers in stone, and their work may have served as patterns for the Chinese to copy from. In conception the designs are evidently not pure Chinese, and I strongly suspect " Mantzu " influence, to use the Chinese term for the aboriginal population.

At Fu-erh-tang there is a particularly fine family temple, and near by a Mantzu cave in an isolated piece of rock. Around many of the mausoleums and family temples ancient stone pillars (wei-tzu, i.e. masts) occur. Wayside shrines and small temples, dedicated to Kwanyin (Goddess of Mercy) and to the tutelary genii are common, the images being carved in stone and mostly coloured blue and white. The day's journey was more than usually interesting; somehow one felt instinctively that one was traversing a region closely associated with man from very ancient times.

Leaving Yuen-fang ch'ang soon after 6 a.m., we traversed country similar to that of the day before, and reached Pai-

(pronounced P'an)-miao ch'ang at 10 o'clock. Here, contrary
to what my map indicated, I found no river. Replies to
inquiries gave it as 30 li farther on, and so it proved. The
map for this region is hopelessly inaccurate, and it was quite
useless attempting to be guided by it. Pai-miao ch'ang is a
small village built on the top of a ridge and surrounded in
part by woods of Cypress and Pine. Crossing an undulating
area we descended by an easy path, finally reaching the
T'ungchiang River, 10 li above Chiangkou. This river is fully
100 yards broad, with red-coloured water and a sluggish
current. Boats were easily secured and we dropped down-
stream to Chiangkou, which we reached at 3 o'clock, just before
a heavy thunderstorm broke. The day's journey was said to
be 70 li, the road was easy, with flora and scenery identical
with that of the preceding days.

Chiangkou (alt. 1600 feet) is the second town in size and
importance in the department of Pa Chou. It consists of
about 500 houses, built on the fringe of a promontory between
two rivers, backed by low, steep, well-wooded hills. The
rivers unite at this point and are navigable downwards to
Chungking. The more easterly stream descends from T'ungchiang Hsien, the westerly stream from Pa Chou, each town
distant from Chiangkou 180 li. Both streams are navigable
for small boats up-stream to these towns.

A Fêng Chou (official next below a Chou in rank) resides
at Chiangkou. From a distance the town looks well-built
and prosperous, but it does not improve on closer inspection.
The position is admirable and undoubtedly the town is of considerable commercial importance, yet we had great difficulty
in exchanging twenty taels of silver. Like other towns we had
passed through, Chiangkou was feeling the suppression of the
opium traffic severely, and until new industries arise to take
the place of the opium trade the resources of all these places
will be crippled.

We found accommodation in a poor but quiet inn, and,
thanks to the thunderstorm, no curious crowd gathered to
annoy us. My principal men spent several hours in finding
out a cross-country road to Yilung Hsien, and eventually
succeeded.

On leaving Chiangkou we ferried across the Pa Chou River and then made a steep ascent of a few hundred feet. The rest of the day we meandered along the crest of a range of low mountains, following an undulating path. In parts the road was good, in others ankle deep in slippery mud. Thundershowers fell at intervals and it was fairly cool.

The country generally is similar to that traversed during previous days. Tobacco is a rather common crop hereabouts and we saw a little cotton. Maize is very rare, but rice is abundantly cultivated. Shrines and small temples continued common and in good repair. Kwanyin and Tuti are the common deities, the latter representing an old man and his wife, constituting the tutelary genii. Dignified, ornately carved tombstones and mausoleums were everywhere in evidence.

Our intended destination for the day was Chên-lung ch'ang, 60 li from Chiangkou, but on reaching there we found market in full swing, and, to avoid the crowd, we journeyed on another 6 li. On market days these villages are impossible, from the foreigner's point of view. I rode through this village in my chair, and the crowd which gathered at the upper end of the place mustered several hundreds. Wine appears to flow freely on market days and many were under its exciting influence. Prudence as well as comfort therefore demands that one avoid all crowds as much as possible when travelling in the interior regions of China. Women attend these markets in force and appear to be a power in this part of the Celestial Empire. Their bearing and manners generally are very free for Chinese women ; natural, unbound feet are the rule.

Chên-lung ch'ang is clustered on the narrow neck of a sandstone ridge, and in common with all such villages boasts a fine village temple. We lodged for the night in a poor wayside inn at Hei-tou-k'an, alt. 3100 feet.

The next day was cool, with showers at odd times, but of no consequence. With the exception of one steep descent and an ascent in the late afternoon, the road was more or less level all day, traversing the tops of the low mountains. These sandstone mountains are dissected by innumerable deep, narrow ravines, clothed with Pine, Cypress, and a dense jungle

THE REDBEAN TREE (ORMOSIA HOSIEI) 60 FT. TALL, GIRTH 20 FT.

of miscellaneous shrubs. Unlike limestone country no bottom-lands are formed, and cultivation is relegated to the higher parts of the ranges. Farmhouses are scattered here, there, and everywhere, but the villages are all situated on the tops of the mountains, most frequently on the divide of a ridge.

Fourteen li from Hei-tou-k'an we passed through the village of Tai-lu ch'ang, where market was in progress and many pigs on sale. Thirty li from this place we passed Ting-shan ch'ang, a village of considerable size, charmingly situated on the neck of a ridge, backed by a Chaitzu and a fine cypress grove. Chaitzu, of which frequent mention has been made, are a feature of these parts. They are old forts, said to have been mostly constructed during the great sectarian rebellion of A.D. 1796–1803. A small official (Hsao-shoa-tang) resides at Ting-shan ch'ang. In spite of its fine situation this village was unusually filthy and was dominated with the strong odours of a wine distillery. The usual crowd of loafers followed us for some distance on quitting this village.

In the late afternoon we arrived at Lung-peh ch'ang, alt. 3000 feet, after travelling 74 li. We lodged in a rambling, dilapidated inn, fairly clean, with rooms removed some little distance from the street—the village sewer. Market not being in progress the crowd of inquisitive idlers was relatively small.

The flora was not particularly interesting, but we passed a number of fine Camphor trees (*Cinnamomum Camphora*). The crops, however, were rich and varied. Rice and sweet potato preponderate, odd patches of cotton were noted and also others of Indigo (*Strobilanthes flaccidifolius*). In the afternoon coolies laden with salt passed us. This salt is pure white and granular and comes from Nanpu Hsien. From our lodging Ting-shan ch'ang was visible, 30 li distant and nearly due east. The map shows a river flowing past this village, but the only one we could get tidings of was 50 li from that place.

After a comfortable night's rest we continued our journey through country similar to that of foregoing days, but less well-wooded and more inclined to be arid, with broader valleys more under cultivation. Our route followed the boundary between Pa Chou and Yilung Hsien. We passed through two market villages and stayed for the night in a farmhouse 1 li before reaching Fu-ling ch'ang, alt. 2800 feet. We purposely stopped at this place in order to escape market day at the village, but did not avoid a constant crowd until after dark, when the doors were closed. We found all these crowds quiet and orderly enough, but a continuous mass of faces, with wooden expression, blocking the doorway, obstructing light and air, is very trying. Immensely useful as these markets are to the country-side, they have decided drawbacks from a traveller's point of view. A good police force is really more necessary in these villages than in the cities. The more lawless element fears a Hsien (Magistrate), but has little respect for a Ti-pao (Village Head-man). Local produce is mostly in evidence in these markets; a few needles, aniline dyes, trumpery odds and ends, chiefly of Japanese origin, are about the only foreign goods met with.

We saw more cotton during the day than we had elsewhere observed on this journey, and the crop looked flourishing. Kao-liang (*Sorghum vulgare*) was a common crop, but rice and sweet potato again preponderated. The sorghum and rice were bursting into ear. Wood Oil trees occur, but are not plentiful, and commercially this crop is unimportant hereabouts. Mixed with the cotton were odd plants of the oil-seed yielding *Sesamum indicum* (" Hsiang-yu ").

In the late afternoon we traversed country which somewhat resembled that around Tunghsiang Hsien—on all sides, as far as the eye could see, nothing but ridge upon ridge of low sandstone mountains. These ranges average about 3000 feet in altitude, those to the east and north being higher than those to the west and south. The map is all wrong for the region, so I could not definitely place our route. The river Sheng-to, so boldly indicated, escaped us, though we should have crossed it had the map been correct. The market villages passed were smaller than heretofore, very filthy and stinking, yet most charmingly situated on the neck of low ridges, and well shaded with trees. Camphor trees are very common, and " Pride of India " trees (*Melia Azedarach*) particularly abundant. The stage said to be 70 li proved very easy, the weather being dull and cool.

Our stay over at the farmhouse was hardly a success;
we had a full crowd until bedtime, and in spite of fair promises
four of my men who remained in the house with me had
neither dinner nor bedding. As a punishment I paid only
half our usual rate, much to the householder's chagrin. Fu-
ling ch'ang was quite deserted when we passed through in the
early morning. It occupies the narrow neck of a sandstone
ridge, after the usual manner of these villages. The same is
true of Shih-ya ch'ang, 30 li farther on. Ten li beyond this
latter village we passed a nine-storied pagoda and sighted the
town of Yilung Hsien, to the northwards, about a mile distant
as the crow flies and at equal altitude (2500 feet). Yilung
is a very small town, situated on the mountain-top, backed
by a steep bluff and surrounded by a wall of dressed sandstone.
Two-thirds of the land enclosed within the city wall is given
over to cultivation. We passed to the south-west of the town
by a road which makes a steep descent and ascent and then
meanders along the tops of the mountains until Tu-mên-pu
is reached. The mountains are lower, more flat, the valleys
wider, and the whole country more treeless. Cotton is abundantly cultivated throughout this region, and it is evident
that the district of Yilung produces a very considerable quantity. Rice and sweet potato are the common crops, the
latter thriving on the hot almost soilless rocks. The earth is
drawn into ridges, often leaving bare rock between, and cuttings
are inserted. These cuttings, leafy shoots about 6 inches long,
quickly take root and form plants that produce an abundant
crop. Sorghum is fairly common in places, but maize is very
scarce. Stone monuments were less in evidence, but we passed
a fine O-mi-to Fu stone surmounted by a hideous T'eng-kou.
Six old hats protected this stone from the rain and sun; in
front was a huge mass of ashes and the remains of many
Joss sticks. We were informed that the tutelary genius of
this spot is renowned for his benevolence, and that it was
hoped shortly to erect a shrine over the spot.

We had been unfortunate in the matter of market days all
along, and found another in progress at Tu-mên-pu. Seemingly
having gained nothing by staying the night a little beyond or
before reaching these villages, we experimented and stayed at

one. It was not a success. A mob rushed our inn and bedlam reigned for a couple of hours. Eventually it thinned down, but many of the more insistent and curious remained until bedtime. There was much noise, but the crowd was friendly enough; nevertheless, I was glad it proved to be the last market village of its kind we encountered before reaching Chengtu.

Tu-mên-pu or ch'ang, alt. 1950 feet, 70 li from Fu-ling ch'ang, is a large and prosperous village boasting much trade on market-days. Something of everything in the way of native produce was on sale, and the narrow street was thronged to overflowing. Five li before reaching this place our road converged with one leading to Pa Chou city by way of Yilung Hsien.

I had a poor night's sleep in consequence of loud talking being carried on far into the early hours, a woman (as usual) being the principal offender. This was an emphatic reminder of the hubbub of the crowd which besieged us on arrival, and I was really glad to quit Tu-mên-pu. A few li beyond this village we branched off from the main road, which goes to Nanpu Hsien. Much salt comes from this township, and during the last two or three days we had met many carriers laden with this commodity.

Forty li beyond Tu-mên-pu we passed the poor village of Shui-kuan-ying, protected by dilapidated gates which denote its former military character. In years gone by it was a barrier of some considerable importance. Twenty li farther on we reached the village of Chin-ya ch'ang, alt. 2150 feet, which differs from all we had met with heretofore in having a broad main street fully exposed to the heavens. To our great joy market was not in progress. We found lodgings in a new and quiet inn, which proved a welcome change ; the people, too, were courteous and much less inquisitive. The day was exceptionally hot, and all were glad to reach the end of the allotted stage of 60 li. Twelve li before reaching Chin-ya ch'ang we struck a main road leading from Nanpu Hsien, and following it entered the village through an isolated ornate gateway. Beyond the village is a bluff of grey sandstone studded with square-mouthed caves. These caves are crude imitations of Mantzu caves, and are of recent origin, and purely Chinese.

The day's journey was through less interesting country than usual. The broad valleys and nearly treeless mountains are all under cultivation. Cotton was again common in the forenoon, but much less so afterwards. This crop looked as flourishing as Chinese cotton usually does. Tobacco is sparingly cultivated. The tobacco leaves are merely sun-dried before using, and the quality is therefore poor. Sweet potato was more plentiful than ever ; the arid sandstone rocks evidently suit this crop. Rice was, of course, everywhere abundant, sorghum common, but maize was very scarce and suffering from drought. The Irish potato is very little cultivated in these parts. Around Tu-mên-pu white-wax is produced in small quantities on the Privet (*Ligustrum lucidum*), but the cultivation is slovenly carried out, the trees being dwarf and ill-cared for. A few Cypress trees were noted, but Paulownia is a common tree, and Wood Oil trees rather plentiful. A little silk is raised, but the industry is unimportant hereabouts. Odd trees of the Banyan (*Ficus infectoria*) occur near houses and shrines. We passed a few fine tombs, but the average headstone is less ornate than those formerly met with.

We experienced a brief but terrific thunderstorm during the early hours of the morning, and rain continued to fall slightly when we set out from Chin-ya ch'ang. For 20 li we followed an abominable road of mud. This was very greasy, and caused many of us to come " croppers." Ultimately, we reached a paved road, and, 6 li farther, a tributary stream of the Kialing River. This tributary is broad, broken by cataracts and rapids, and quite unnavigable at this point. It unites with the Kialing River, locally known as the " Paoning Ho," at Ho-che kuan. This is a small riverine port boasting a remarkably fine shop where coal, lime, and especially Chinese wine (sam-shu), were on sale. On the paved road we met several men carrying Bombay cotton yarn—the first example of foreign goods we had encountered on the whole journey !

At Ho-che kuan the Kialing River is smooth and placid, and when in flood is fully 400 yards broad. We ferried across the river to the right bank, and then traversed an alluvial flat of considerable size, highly cultivated with rice and sorghum, with here and there a little abutilon hemp. At the head of this flat, some 10 li from the river, we crossed over some levelled hillocks into a basin—evidently an old lake bed—surrounded by bare mountains 200 to 500 feet high. This depression was a lake of luxuriant padi (rice), with houses here and there, nestling in clumps of trees. From this basin we passed through a low, narrow gap between the hills, and came abruptly to the Paoning River a little below the city itself. We were ferried across and found lodgings in a large and fairly comfortable inn. The flora of the day's journey was without special interest, Cypress being the only kind of tree really common. But shading some graves, opposite Ho-che kuan, occurs the largest specimen of the " Pride of India " (*Melia Azedarach*) I have met with. This tree is 70 feet tall, and 10 feet in girth.

Paoning Fu is a city of past rather than of present greatness. It is still a most important administrative centre, but its real interest lies in its great historic past. From the early days of Chinese conquest it has been a strategical point of vast importance. During the Ming dynasty (A.D. 1368–1644) a generalissimo of forces had a palace here. The terrible rebel, Chang Hien-tsung (A.D. 1630–46 *circa*), ravaged the country roundabout, but spared the city itself. The result is that many of the official residences and temples date back to ancient times.

Formerly Paoning was the centre of a lucrative and thriving silk industry, but this has steadily declined during the last twenty years, and to-day it is a mere figment in comparison. Attempts are now being made by the officials to rejuvenate and foster this industry, which apparently failed more through lack of business ability and tenacity than anything else. On the neighbouring hills I was told " wild silk " is produced, the " worms " feeding on the leaves of a scrub Oak, " Ching-kang " (*Quercus serrata*).

The city occupies an extensive alluvial flat on the left bank of the river within an amphitheatre of low, bare, often pyramidal, hills, 300 to 600 feet high. Viewed from the opposite bank there are no outstanding architectural features visible, save a pavilion, which is practically the only building breaking

NAN-CHING KUAN, OPPOSITE PAONING FU

the monotony of level roofs. The area within the city walls is largely occupied by yamêns, temples, and residences of the more wealthy. Business is mostly carried on outside the city proper, and is confined mainly to one street. Umbrellas were the most noticeable articles on sale, but the city is famous for its superior vinegar, great jars of which were on view.

Hedges of the thorny shrub, *Citrus trifoliata*, are a prominent feature of this city and its suburbs, giving to the quieter streets a country-lane-like appearance. The water supply of the city is from wells, which are often very deep. This water is said to be good, but that supplied to our inn had a very " earthy " flavour. From what I saw of the city during a day's stay there, I received the impression of its being clean, its people very orderly and courteous, and the decline in its prosperity most marked. The Paoning Ho is a shallow river, and opposite the city about 500 yards broad when in flood. It is navigable for boats of considerable size downwards to Chungking. Up-stream small boats ascend to Kuangyuan Hsien. A certain amount of merchandise descends in small boats from Pikou, in Kansu, to Chaohua Hsien. These rivers are most important to Paoning Fu, for, in addition to export trade, the coal and wood used in the city itself are conveyed over these waterways. On the right bank facing the city is a ledge of cliff, on which nestle several temples and pavilions, sheltered by groves of Cypress. In a gap in this cliff is situated the busy little village of Nan-ching kuan. Timber is very scarce around Paoning. Cypress wood is commonly used in house-building ; Alder wood (*Alnus cremastogyne*) occasionally being employed for window frames, etc., but its chief use is as fuel. Pine occurs, but, save as fuel, is worthless. Cunninghamia, that most useful of Chinese conifers, does not occur in this neighbourhood. The wood of the Hung-tou tree (*Ormosia Hosiei*), so highly esteemed for carpentry, was formerly fairly common and cheap. To-day, however, it has to be brought from a distance, and, in consequence, is expensive. Oak and " Huang-lien " (*Pistacia chinensis*) are the only other timber trees of note. Paoning is an important missionary centre, and the seat of a Protestant bishopric. During my brief visit I had the pleasure of spending a few hours with the kindly and energetic Bishop Cassels and certain of his coadjutors, who did all they could to render my stay pleasant.

Leaving Paoning Fu and following the main road via Tungchuan Fu, by easy stages I entered the city of Chengtu Fu nine days later, having occupied fifty-four days on the journey from Ichang.

The journey from Tunghsiang Hsien to Paoning Fu fully bore out my expectations. The crowds on market-days were a decided drawback, but not once was I insulted or called (in my hearing) uncomplimentary names. The avaricious greed and cunning of the inhabitants were most marked. They were constantly putting up the prices of food-stuffs on my followers, which led to much argument and high words, and several times I was called upon to settle such disputes. The greed of the Szechuanese peasant and small shopkeeper is a byword among the Chinese of other provinces. The term "Szechuan Lao-ssu" ("Szechuan Rat") is applied derisively to the whole population by the Chinese from other provinces. Niggardly and avaricious they undoubtedly are, but they are great agriculturists, and the question of the "mote and beam" may well be left open. As mentioned before, the province is largely peopled by descendants of immigrants, and these folk almost invariably style themselves men of the provinces their ancestors came from !

The outstanding features of this ancient part of Szechuan are :—

1. The elaborate system of market villages situated at equal distances of 30 li apart, each with its nine market-days per month, and alternating with the markets of neighbouring villages. Each village is situated on the mountain-top and usually on the neck of a divide, with one central more or less covered street.

2. The rice belt is confined to the mountain slopes and summits, the valleys being ravines, jungle-clad as a rule, with little or no cultivatable bottom-lands. The highly cultivated nature of the region and the presence of cotton in quantity around Yilung Hsien.

3. The numerous fine mausoleums with remarkably good

sculpturing ; the peculiar, dignified style of headstones and mural monuments generally. The number of wayside shrines and deities all in excellent repair.

4. The independent bearing and buxom appearance of the women, and their evident influence in general market business. Throughout the whole region natural, unbound feet are the rule.

5. The region is far from being thickly populated, and cannot be termed wealthy, but apparently it is largely self-contained and self-sufficient.

6. The intense curiosity of the people due to the fact that few had ever seen a foreigner before.

CHAPTER IX

THE CHENGTU PLAIN

" The Garden of Western China "

THE plain of Chengtu is the only large expanse of level ground in the great province of Szechuan; it is also one of the richest, most fertile, and thickly populated areas in the whole of China. Its extreme length from Chiang-kou in the south to Hsao-shui Ho beyond Mienchu Hsien in the north, is about 80 miles as the crow flies; its extreme width from Chao-chia-tu in the east to Kuan Hsien in the west, about 65 miles, in a straight line. From Kiung Chou in the extreme south-west to its north-east limits beyond Teyang Hsien is about 80 miles. The circumferential boundaries are very irregular, the total area being under 3500 square miles. Chengtu Fu, the provincial capital, and seventeen other walled cities, are situated on this plain, together with very many un-walled towns of large size. Farmhouses dot the plain in every direction; the total population probably exceeds 6,000,000.

This plain is really part basin part sloping alluvial delta, having an elevation ranging from about 1500 feet above sea-level in the south and east to 2300 feet in the north-west and west. It is bounded to the west and north-west by the steep descent of a high mountainous region, which at very little distance from it reaches above the snowline. In the extreme north-west the snowclad Chiuting shan actually overlooks the plain. On its other boundaries the sandstone hills of the Red Basin rise sharply in bluffs 1000 to 1500 feet above the level of the plain. The high barrier ranges protect the plain from the cold northerly and westerly winds, but to these must be ascribed the rapid changes in temperature, the fogs, raw atmosphere, and the overcast skies so characteristic of Chengtu Fu.

The plain owes its abundant fertility to a complete and marvellous system of irrigation, inaugurated some 2100 years ago by a Chinese official named Li-ping and his son. The headquarters of this irrigation system is Kuan Hsien, a city situated on the extreme western edge of the plain, where the Min River debouches from the mountains. The principle on which the system is based is simple in conception, but very intricate in detail. An obstructing hill called Li-tiu shan was first cut through for the purpose of leading the waters through and distributing them over the plain. The passage having been excavated, the waters of the Min River were divided, by means of an inverted V-shaped dyke, a little distance above the canal into two main streams, the " South " and " North " Rivers, as they are called. The waters of the " North " stream are carried through the Li-tiu shan cut, and after passing through the city of Kuan Hsien are divided into three principal streams. The most southerly of the three, called " The Walking Horse," flows directly east, and irrigates the districts of Pi Hsien and Chengtu. The central stream, called the " Cedar Stem River," flows north-east, and is utilized to irrigate the western and northern parts of the above-named districts. Branches of these two streams flow past the south and north walls of Chengtu, uniting near the east gate of the city. The third, or northern branch, known as the " South Rush River," flows north towards the city of Pêng Hsien, and then south-eastwards past Han Chou. All the subdivisions of this branch and its anastomosing canals and ditches unite near Chao-chia-tu to form the head-waters of the To River, which flows due south past the famous salt-wells of Tzu-liu-ching, and finally enters the Yangtsze at Lu Chou. This " South Rush River " is fed by numerous torrents which descend from the ranges bounding the north-west edge of the plain. These streams—broad, stony, irresponsible things with no defined banks—exist only during rains or the melting of the snow in spring. In crossing the northern parts of the plain the traveller can form some estimate of what the whole was like before the irrigation canals were

dug and dikes erected.   But to return to the system at Kuan
Hsien.   The "South" River, which occupies the original bed
of the Min River, is divided into four principal streams almost
immediately opposite the Li-tiu Hill.   The most easterly
branch, named the "Peaceful River," irrigates the districts
of Kuan Hsien, Pi Hsien, and Shuangliu Hsien.   The next
branch, called the "Sheep Horse River," irrigates other parts
of the above-named districts, uniting with the "Peaceful
River," at Hsinhsin Hsien.   The third stream, called "Black
Stone River," irrigates the department of Chungching Chou,
and unites with the other streams at Hsinhsin Hsien.   The
fourth stream, called "Sand Ditch," flows south-west through
Tayi Hsien and Kiung Chou, joining the other streams at
Hsinhsin Hsien.   All the streams which intersect the Chengtu
Plain, save those forming the upper waters of the To River,
unite at Chiangkou, a village at the extreme south-eastern edge
of the plain, some 45 English miles south of Chengtu city.

This system of anastomosing canals, ditches, artificial and
natural streams, forms a complex yet perfect network.   The
current in all is steady and swift, the bunding secure, and
floods unknown.   Not only are all these streams and canals
available for irrigation, but they are also utilized to generate
power required in various industries.   Flour-mills abound,
driven by vertical or horizontally fixed water-wheels.   Similar
mills are used for crushing Chinese rape-seed, preparatory to
pressing for the extraction of the oil.

It must not be supposed that Li-ping and his son completed
the system which obtains to-day.   They were the originators,
and the lines they laid down have been followed and enlarged
upon by succeeding generations.   These famous irrigation
works are perhaps the only public works in all China that are
kept in constant and thorough repair.   Every year the bunding
is repaired and all silt removed from the bed of the channels.
An official styled Shui-li Fu—"Prefect of Water-Ways"—
residing at Chengtu, has charge of the system.   In late winter
the water is diverted at Kuan Hsien from the "North" River
to admit of the removing of silt, etc.   In the early spring,
conducted with much pomp, there is an annual ceremony of
turning on the waters.   The motto of Li-ping, " Shen tao t'an,

THE DIVIDED WATERS AND BRIDGE (AN-LAN CHIAO) 250 YDS. LONG

CHANNEL CUT THROUGH THE LI-TIU SHAN BY LI-PING

ti tso yen '' (Dig the bed deep, keep the banks low), has become an established law in these parts, and is rigorously carried into effect. Amidst so much that is decaying and corrupt in China it is refreshing to find an old institution maintaining its standard of excellence and usefulness through century after century. The originators of this work have been deified, and two magnificent temples overlooking their work at Kuan Hsien bear witness to the gratitude of the millions who have enjoyed, and continue to enjoy, prosperity from the labours of the famous Li-ping and his son. The " hero-worship " here exemplified would do credit to the people of any land.

The larger of the two temples merits some description. It is by far the finest example I have seen in my travels, and is probably not excelled by any temple in all China. It nestles midst a grove of fine trees, facing the river on the side of a hill, with broad flights of steps leading from terrace to terrace. The buildings are of wood, finely carved and lacquered. The court-yards of stone are broad and spacious, with ornaments in bronze and iron of old and unique workmanship. There are figures representing Li-ping, his wife and son, also many finely gilded and inscribed votive boards, gifts of a long line of succeeding emperors, viceroys, gentry, and guilds. Not a weed is allowed to grow, the whole place being kept scrupulously clean by the Taouist priests in charge. In the courtyards are many interesting trees and shrubs, trained in Chinese manner with consummate skill. Two magnificent specimens of the Crêpe Myrtle (*Lagerstrœmia indica*), trained into the shape of a fan some 25 feet high by 12 feet wide, and said to be over 200 years old, are finer than anything of the kind I have seen elsewhere.

The whole of the plain is subdivided into small fields, every field or series of fields having its own level, differing (sometimes only by one or two inches) from that of its neighbours. This arrangement necessitates a complicated code of regulations, which, sanctioned by custom and usage, determines the pro-portions in which the water of any one canal is distributed into its branches, and the order of succession in which proprietors of different fields are allowed to make use of it. The system has been so far perfected that each rice field receives, exactly at the right time, a sufficient supply of running water. So complete is the whole arrangement that scarcity, much less famine, is practically unknown on the Chengtu Plain.

There are no extremes of climate in this region. In summer the temperature seldom reaches 100° F., in the shade ; in winter it seldom falls below 35° F. It is humid at all times and essentially cloudy, more especially in winter, when the sun is rarely seen, owing to banks of mists. The land is always under cultivation, yielding two main crops that ripen in April or May, and August or September respectively. Catch crops are obtained between these two main harvests. Rice is the chief summer crop, but certain districts produce millet, sugar, pulse, Indigo (*Strobilanthes flaccidifolius*), and tobacco in quantity, Pi Hsien being noted in particular for the latter crop. Wheat and Chinese rape are the chief winter crops with Broadbeans (*Vicia Faba*), peas, barley, and Hemp (*Cannabis sativa*), common in certain districts. Wên-chiang Hsien is famous for its hemp, which is grown in quantity as a winter crop and exported largely to other parts of Szechuan and down river. This product, known colloquially as " Huo-ma," has been wrongly identified by many travellers. As summer crops, Ramie or " Hsien-ma " (*Bœhmeria nivea*) and Abutilon hemp or " Tuen-ma " (*Abutilon Avicennæ*) are both cultivated more or less in quantity. The only Jute or " Huang-ma " (*Corchorus capsularis*) I ever saw was in July 1910, growing near Yao-chia-tu. In the northern parts of the plain, Mienchu and Teyang Hsiens, a little cotton is raised, but commercially the crop is unimportant. Opium was never cultivated in quantity on the plain.

All the Chinese vegetables and culinary oil-producing plants are cultivated in quantity in the Chengtu Plain, and their general excellence is not excelled elsewhere. To enumerate them it would be necessary to give a complete list of such plants cultivated in all but the coldest parts of China. This enumeration is reserved for a subsequent chapter.

A striking feature of the plain is the enormous number of large houses and farmsteads dotted here, there, and everywhere, and shaded by groves of Bamboo, Nanmu, and Cypress. The frequency of these houses, with their enveloping groves, gives a well-wooded appearance to the entire region, and the general view is broken up in such a manner that from no point can many miles of the plain be seen at one time.

The variety of trees is very great ; fully fifty species could easily be enumerated. Alongside the streams and ditches, Alder, " Ching-mu " (*Alnus cremastogyne*) abounds, and forms one of the principal sources of fuel. In the more northern parts of the plain the curious *Camptotheca acuminata*, with clean trunk, grey bark, and globose heads of small white flowers, displaces the Alder. Around the houses Bamboo, Oak, "Pride of India," Soap trees (*Gleditsia*), Cypress, and Nanmu are the commonest trees. The Nanmu is a special feature around temples. Several species of the genus *Machilus* are called Nanmu, all agreeing in being stately, tall, umbrageous evergreens. The wood they yield is highly valued, and the trees are particularly handsome. The Banyan tree, so abundant a little farther south, is very rare here, and neither Pine nor Chinese Fir (*Cunninghamia*) are common. Occasionally trees of the Red Bean (*Ormosia Hosiei*) occur, always, however, in temple-yards or shading wayside shrines. The great industry of Chengtu Fu is sericulture, consequently Mulberry trees abound, and *Cudrania tricuspidata* (Tsa shu), the leaves of which are also used for feeding silkworms, is likewise fairly common.

In such a highly cultivated area the natural flora has, of course, been destroyed. The few indigenous shrubs and herbs that remain are relegated to the sides of streams and graveyards. In places the Chinese Pampas Grass (*Miscanthus sinensis* and *M. latifolius*) is common ; in autumn the fawn-coloured plumes are most attractive. Occasionally thorny shrubs like Barberry, Christ's thorn, colloquially " Teh-li-pê kuo-tzu " (*Paliurus ramosissimus*), and " San-chia pi " (*Acanthopanax aculeatum*) are used as hedge plants. The commonest fence, however, is made by bending down and interlacing the bamboo-culms.

Since the plain is strewn with cities, villages, and farmsteads, a network of roadways necessarily obtains. A main artery extends north-north-east, through the plain and beyond to Shensi province, and ultimately reaches far-distant Peking. This road was commenced from the Shensi end by the great

Shih Hwang-ti (he who commenced building the Great Wall) about 220 B.C. It extends from Chengtu in a south-westerly direction to Kiung Chou, and thence to remote Lhassa. Other highways connect the provincial capital with Chungking, the great mart on the Yangtsze River to the south-east ; Kuan Hsien in the west, and the Marches of the Mantzu beyond. Roads of secondary importance link these highways with other roads and connect the capital with all the principal cities of the plain and regions beyond. Most of the roads were originally paved with one or two slabs of stone laid lengthwise down the middle, with bare earth on either side. The constant wheel-barrow traffic, a feature of the entire region, has worn deep grooves into these slabs. All too frequently the slabs have disappeared altogether, leaving unpaved long stretches of roadway. In dry weather these roads are dusty, but easy to travel ; in wet weather they are from ankle to knee-deep in sheer mud. Often they are practically impassable, and travelling over them in ordinary rainy weather is an experience beyond words to describe. They illustrate admirably the contrariety of things which obtain in China generally. Here in the wealthiest region of the west, if not of the whole of China, the average road is of the meanest width, and in an abominable state of repair. There is much talk of the need of railways in China,—true, they are needed badly, but good highways, *roads*, are an infinitely greater want. The highways and byways on the Chengtu Plain are a disgrace to the entire population of this fertile, wealthy region. " What is everybody's business is nobody's business " is a saying that is as applicable in China as in Western lands. The roads exist for the good and welfare of all, but it is nobody's real business to protect them ; they are, in consequence, neglected by all— peasants, farmers, officials, and gentry alike.

Mean as these roadways are, they are spanned by hundreds of large honorary portals and memorial arches, mostly constructed of red, or more rarely grey, sandstone, or occasionally of wood. In the vicinity of the more wealthy cities (Han Chou, for example) these portals and arches are extraordinarily abundant. Many are masterpieces of Chinese architecture. All are well built and covered with sculptures in relief, re-

VIEW IN THE MANCHU SECTION OF CHENGTU CITY

THE VILLAGE TEMPLE, KUNG-CHING CH'ÀNG

presenting scenes of mythical or everyday life. The ends of the ridge pole and the gable eaves are usually long drawn out and revolutely upturned, adding additional lightness and beauty to the whole. These long, exaggerated, upturned eaves are a characteristic feature of the houses, temples, and shrines met with all over this region.

The innumerable ditches, canals, and streams are all well bridged. The bridges are kept in good repair, and reflect the highest credit on the engineers who constructed them. They are built of red or grey sandstone, more rarely of wood, as near Han Chou. The stone bridges vary from one to a dozen or more arches, sometimes hog-backed, but more usually the " Roman Arch " is employed ; others are of causeway or trestle design, with or without balustrades, ranging from a single slab laid across a narrow ditch to many such laid on a series of piers built in the bed of the streams. Near Sintu Hsien there is an example of one of these trestle or pier-bridges 120 yards long. Outside the east gate of Chengtu is a red-sandstone bridge of nine arches, which is generally regarded as the bridge mentioned by Marco Polo. A similar bridge exists near Yao-chia-tu, but this has some twenty arches. Immediately outside Han Chou there is a covered wooden bridge, 120 yards long, 6 yards broad, resting on eight stone piers. This bridge, known as the Chin-ying chiao (Bridge of the Golden Goose), is the handsomest, most ornate wooden structure of its kind I have met with in my travels.

In reference to the bunding of the streams and canals it should be mentioned that cobble-stones enclosed within long sausage-shaped, bamboo-latticed crates are universally employed for this purpose. This system is said to date back to the later times of the Ming dynasty only. Previous to that period the principal abutments and revetments were of iron, fashioned into the shape of gigantic oxen, turtles, pillars, etc. At the places where canals unite or divide, or where the water cascades to a lower level, the earthworks are protected by walls of stones firmly cemented together.

Another item, and one which astonishes every traveller, is the enormous size of the blocks of stone used in the bridges, more especially those erected on piers. I have no exact measurements by me, but these slabs would average at least 12 yards long by 20 inches square at the ends. Commonly the blocks are of hard limestone, occasionally of conglomerate. The slabs of sandstone when used are shorter. At Chao-chia-tu sandstone slabs are used as fencing.

Any attempt to describe the cities on the Chengtu Plain would necessitate more space than is at my disposal. They differ, with the exception of the provincial capital, in no marked particular from other cities of Szechuan. In size they vary considerably, some of the large unwalled towns being commercially more important than the walled cities. Most of the cities and surrounding districts are noted for certain things ; for example, Mienchu Hsien for its wheaten flour and paper, P'i Hsien for tobacco, Wên-chiang Hsien for hemp, P'êng Hsien for indigo, Shuangliu Hsien for straw-braid, and so on. The majority of these cities are very ancient ; all contain fine temples, as becomes such centres of wealth. Chengtu (long. 104° 2' E., lat. 30° 38' N.) was described by Marco Polo, who visited it during the thirteenth century, as a " rich and noble city." Modern travellers, and their name is well-nigh legion, have all agreed with the great Venetian's dictum. In many respects Chengtu, with its population of 350,000 people, is probably the finest city in the whole of China. It is built on a totally different plan from that of Peking, or even Canton, so that comparisons are difficult. The present city of Chengtu is comparatively modern, but occupies much the same site as the capital of the aboriginal kingdom of Shu. This kingdom was conquered by Shih Hwang-ti (the " First Emperor ") some time between 221–209 B.C., who nominally added it to his dominions. The succeeding dynasty of Han (206 B.C. to A.D. 25) incorporated it as an integral part of China. During the epoch of the Three Kingdoms the site (or thereabouts) of the city was occupied as the capital of the kingdom under Liu-pei. Succeeding dynasties have always made it a most important seat of administration, and princes of the imperial clan or viceroys have resided there. It is still the seat of a Viceroy who governs the province of Szechuan and nominally controls all Thibetan affairs.

Great Britain, France, and Germany have each established

a Consulate-General there, but on the plea that the city is not an " open port," the Chinese have successfully resisted the purchase of land on which to erect suitable houses and offices for the staffs representing these Powers. The result is that these officers are housed in dilapidated Chinese quarters, insanitary, dangerous to health, and unbecoming the dignity of the Powers they represent. It is nothing short of a scandal to thrust men into such abominable quarters. Chengtu Fu is far removed from London, Paris, and Berlin, also from Peking, but is it fitting to make backwoodsmen of these representatives ? Missionaries of every denomination are firmly entrenched at Chengtu, and can acquire all the property their funds admit of either for residences, hospitals, schools, or churches.

The city is surrounded by a magnificent wall, some 9 miles in circumference, with eight bastions, pierced by four fine gates. This wall is 66 feet broad at base, 35 feet high, and 40 feet broad at top, along which runs a crenulated balustrade. It is faced and paved with hard brick (the walls of all the other cities on the plain are of sandstone), and is kept in thorough repair. During Manchu times a Tartar garrison was stationed here, a large area on the south-west side of the city being walled off to form a Manchu city. Within the city walls are many fine residences, private and official, temples, a large parade ground, etc. The city is clean and orderly, with an efficient police. To wander through the streets noting the varied industries carried on is a liberal education in Chinese ways of doing things. The wares on sale are of infinite variety, and are themselves indicative of the wealth which is everywhere apparent. The shop-signs, lacquered and gilded, hang vertically downwards, and proclaim in their large artistic characters the titles of the shops and the wares on sale. The city is full of officials, both in and out of office, who move about the streets in sedan-chairs carried at a great speed. The chairs are peculiar in having the long poles curved, with the body of the chair resting on top of the curve. When carried, such a chair is well above the heads of the crowd. The streets are always crowded with pedestrians, chairs, and wheel-barrows. Different trades occupy their own particular quarter. Certain streets are

devoted to carpentry in all its branches, boot-shops, shops
devoted to hornware, skins and furs, embroideries, second-hand
clothes shops, silk goods, foreign goods, and so forth. Silk-
weaving is the great industry in Chengtu, hundreds of looms
being in use.

Evidences of Occidental influence abound. A provincial
university and many schools for imparting Western learning
exist. Two agricultural experimental farms, an arsenal, mint,
bazaar, and many buildings of semi-foreign design. The arsenal
and farms are outside the city. An electric lighting plant was
operating at the time of my last visit (1910), and the installation
of a telephone service was in progress. The Imperial Post
is strongly established here under control of Europeans, and
this is the only Western innovation really accomplishing good
work. The others (and I have not covered them all) are
experiments pure and simple. These are controlled by
officials among whom jealousy is rife and peculation not unknown. The good intentions of honest officials are easily
nullified by jealous-minded sycophants and ultra-conserva-
tives. The city exhibits numerous examples of blighted
experiments, some of them mere follies, but the majority
calculated to be beneficial if properly controlled and carried
through. The city-fathers and officials have exhibited mad
haste to acquire such Western knowledge as they deem useful.
They have no real idea of what they want, and there is little
co-ordination in any matter. The students rule the colleges ;
their fathers, the gentry, rule the province. " China for the
Chinese," and "away with all foreigners and foreign influence"
is their slogan. This cry is perfectly legitimate, but they should
move slowly. They think they are fully fledged men, whereas
they are mere babes in the knowledge of the things they covet
so much. The unfortunate Rebellion which has spread with
such rapidity and brought about so much disaster to the nation,
originated with the hot-heads of Chengtu. Primarily it was
aimed not so much against the dynasty as against foreign capital.
The Central Government had agreed to a foreign loan, which,
amongst other things, had for its object the construction of a
railway from Hankow to Chungking. It was this loan that was
the fat in the fire which produced the conflagration—the last

A MEMORIAL ARCH

straw, if you will, but the primary cause of the Rebellion. The dynasty has been dethroned (it was effete, anyway, and should have passed fifty years ago), a dictatorship under the guise of a republic cleverly formed by the only man who can save China from anarchy if not disruption—Yuan-shih-kai. But foreign loans have become more absolutely necessary than ever before. The present system of government can only be transient, another dynasty must arise. I mentioned above that the province was under a viceroy, and that the gentry ruled the province. This is the keynote to the whole difficulty. The Viceroy had to carry out the instructions of the Imperial Government at Peking ; he had also to please the gentry. The wishes of the two powers became diametrically opposed, and not all the tact of the cleverest diplomats could save the situation. The Viceroy (Chao Êrh-hsün) was removed to Manchuria, and his brother (Chao Êrh-fêng), recalled from the Thibetan Marches (where China's new toy, in the shape of an army modelled on quasi-Western lines, had been indulging in an altogether uncalled-for war of aggression), appointed to the post. The new Viceroy arrived too late to check the revolt, and was ultimately murdered. The gentry have declared that no foreign capital, and the necessary foreign supervision of such capital, shall enter into the construction of a railway in Szechuan. With Chinese money and Chinese engineers the scheme shall be accomplished, say these autocrats. The Central Government thought otherwise and made other arrangements. Then came the revolt, fulminated by the gentry of the Chengtu Plain, which speedily got beyond their control, and where it will really end is beyond prophecy. The Manchu dynasty, when it ascended the Dragon throne in A.D. 1644, immediately set to work and rescued Szechuan from the bloody grip of the rebel and arch-destroyer, Chang Hien-tsung, and brought peace to the land. Two hundred and sixty-seven years later this dynasty has been dethroned by rebellion initiated by the gentry of the Chengtu Plain. Dynasties and republics may come and go, but in the future, as in the past, industry, combined with agricultural skill, will continue to win sustenance, derive wealth, influence, and power from this fertile and beautiful region—the Garden of Western China.

CHAPTER X

NORTH-WESTERN SZECHUAN

NARRATIVE OF A CROSS-MOUNTAIN JOURNEY TO SUNGPAN TING

A FEW days after our arrival at Chengtu in 1910 I determined upon a journey to the border-town of Sungpan Ting, for the express purpose of securing seeds and herbarium specimens of certain new coniferous trees previously discovered by me in that region. During 1903 and again in 1904 I had visited this interesting town. On the first occasion I travelled by the ordinary main road, via Kuan Hsien and the Min Valley. The next year I followed the great north road across the Plain of Chengtu to Mien Chou, then travelled via Chungpa and Lungan Fu, by another recognized highway. On these journeys I gleaned tidings of a by-road leading from Shihch'uan Hsien across the mountains, finally connecting with both the above routes. This route promised to be interesting as well as novel. Only Roman Catholic missionaries had previously traversed it, so far as I could learn. An Hsien was selected as the real starting-point for this trip.

With this object in view we passed through the north gate of Chengtu city early on the morning of 8th August. Following the north road as far as the city of Han Chou, then branching off and travelling via Shihfang Hsien and Mienchu Hsien, we reached the city of An Hsien, some 300 li from Chengtu, after three and a half days. The road led us right through the luxuriant Chengtu Plain to its extreme north-western limits near Hsao-shui Ho. Afterwards we crossed some low foot-hills to a small stream leading to An Hsien. The journey was very easy, though fatiguing owing to the extreme heat of the season.

The city of An Hsien is small, of little importance, prettily situated on the left bank of a stream backed by bare mountains which rear themselves some 2000 feet above the level of the river. Two streams uniting here form a river navigable during high-water season to Mien Chou, a city on the Fou Ho—the western branch of the Kialing River system. An Hsien is a little beyond the north-western limits of the Chengtu Plain, and its river gives it direct communication with Chungking, during the summer at least.

Leaving by the north gate, we took a road that ascends the main branch of the river which is kept from flooding the city by a well-made low bund of stone slabs, firmly cemented together. After traversing a small cultivated valley we plunged into a rocky defile and crossed the river by an iron suspension bridge, 110 yards long. This bridge is old and in poor repair, and it swayed considerably as we walked singly across. A few miles farther on we recrossed the stream by a similar bridge, and reached Lei-ku-ping, our destination for the day, at 6 p.m. A certain amount of rice is cultivated hereabouts, but maize is the staple crop. As an under-crop to maize, *Amorphophallus konjac* ("Mo-yu") is commonly cultivated, the tubers being used as food after their acrid properties have been removed by washing in water. We met considerable traffic, mostly coolies laden with sheep-skins and medicines from Sungpan, which they put on boats at An Hsien for conveyance to Chungking; much potash (lye) in small tubs, and oil-cakes consisting of the residue of Chinese rape-seed after the oil is expressed. Coal of very poor quality, mostly dust, is obtained in the surrounding mountains, and we met scores of mules, ponies, and coolies engaged in transporting it.

Lei-ku-ping, alt. 2750 feet, is a large market village, possessing one principal street with gates at each end, which are closed after sunset. The centre of a large and important industry in tea, Lei-ku-ping largely supplies Sungpan Ting and the country beyond. The tea is grown in the surrounding districts and brought to the village for sale. Later we shall have more to say concerning this industry.

It rained heavily during the early hours of the morning, and though it was fair when we set out, showers fell the whole forenoon. On leaving Lei-ku-ping we ascended a few hundred feet to the head of a low divide, and then descended to the village of Che-shan, situated on the right bank of a considerable stream. This village shares in the tea industry for the Sung-pan market, but is of less importance than Lei-ku-ping.

From Che-shan to Shihch'uan Hsien the road ascends the right bank of the river, which flows between steep precipitous mountains. The path is usually several hundred feet above the stream, broad and fairly easy for the most part, but constantly ascending and descending. The mountain-sides are steep but, where not absolutely vertical, are all under cultivation, Maize being the staple crop. There is very little limestone, the rocks being chiefly loose sandstone and mud shales. These shales weather rapidly, and the steepest cultivated slopes are usually composed of these rocks.

The river is broad, and could easily be made navigable for boats during the high-water season. Even in its present condition rafts could be floated down, but we saw no traffic whatsoever on its waters. The water was dirty, and much driftwood was strewn along the shores. This is collected, dried, and stacked, forming apparently the principal source of fuel. Trees are very scarce, but around houses occur Sophora, Pistacia, Pteroceltis, *Sterculia platanifolia* (Wu-tung), *Kœlreuteria bipinnata*, and Alder. The Kœlreuteria was just coming into flower ; the flowers are golden yellow produced in large, much-branched, erect panicles ; the leaves are very large and much divided. Shrubs are not plentiful but, much to my surprise, the Tea Rose (*Rosa indica*) is quite common, and evidently spontaneous, by the wayside, on the cliffs, and by the side of the stream.

A few li below the city of Shihch'uan Hsien the river is spanned by a bamboo suspension bridge, about 80 yards long, supported on cables made of split, bamboo culms plaited together. These cables, eight in number, are nearly 1 foot in diameter, and are fastened to stanchions fixed on either side of the river. Two similar cables on either side of the bridge are carried across at higher levels, and have attachments of bamboo rope supporting those which form the base of the structure. A capstan arrangement is used for making the cables taut, and the lower ones are covered with stout wickerwork to form a footway. Like all such structures, this bridge is heavy, sags very much in the middle, and is very unsteady to walk across. The life of these bridges is only a few years, and strong winds often make them very unsafe.

Shihch'uan Hsien is a small city charmingly situated at an altitude of 2800 feet, on the left bank immediately below the junction of two rivers. It is surrounded on all sides by steep, more or less cultivated mountains. Inside the city are many trees, which add considerably to the effect. A pavilion and a small pagoda crown two prominent hills, and assure the " luck " of the place. A narrow suburb runs ribbon-like between the river and the city wall. This wall is broken down in places, and the gates are low and small. We found accommodation in a large, curiously constructed inn remarkable for the strength of its stinks and the abundance of vermin and mosquitoes it sheltered. The day's journey was given as 65 li, but the li were long, consequently the coolies with their loads arrived late. Cash was needed, but on opening a box to obtain some silver for exchange we found that some one had stolen from it about 30 taels and 5 dollars. The load belonged to a coolie we had engaged at Taning Hsien, and retained because he had given unusual satisfaction! The previous day he had engaged a local coolie to carry his load, on the ground that he was feeling sick. He was last seen near Che-shan, still unable to carry his load. Evidently he was the culprit, but he was thoughtful enough to leave us about half the amount contained in the box. Since he had about three-quarters of a day's start I concluded it was best to quietly cut the loss, my first and last in China. The delays incident upon lodging a complaint with the official would have involved me in further expense and trouble, with but small chance of recovering the money lost.

The main road to Sungpan continues to ascend the right bank of the river to its source, then crosses over a range and enters the upper Min Valley at Mao Chou. I had been over most of this route in 1908 when crossing the Chiuting range from near Mienchu Hsien to Tu-mên, thence to Mao Chou. The route we had in view leads to the north-west from Shihch'uan Hsien. From Chengtu to this point we had travelled without escort, but with the difficulties of an unknown route before us

I thought it best to secure such at this city.   Sending my card
to the Hsien's yamên, in the ordinary way, I informed this
official of my project, and asked for the customary escort.   Half
an hour afterwards my card was returned with the information
that there was trouble at Sungpan and no escort would be
supplied !   The refusal was as curt as it was insolent, but
whether the Hsien was actually responsible I never found out.
In the whole of my eleven years' travel in China this was the
first and last experience of official discourtesy.   Thus two
annoying experiences, both unique in their way, yet, happily,
trivial and unimportant, marked my visit to Shihch'uan Hsien,
a town which, from the commencement of my travels in the
western Szechuan, I always had a keen desire to visit.

The next day we left Shihch'uan Hsien at sunrise, glad to
escape from the malodorous, vermin-infested inn.   No one put
in an appearance from the yamên, and no attempt to prevent
our taking the route proposed was made.   I had rather feared
this might happen, but my fears were fortunately groundless.
On leaving the city by the north gate we struck a stream nearly
equal in volume to the main river.   The road ascends the left
bank, and almost immediately plunges into a narrow, wild
ravine, through which we continued the whole day.   Like all
such roads it skirts the mountain-side, being usually several
hundred feet above the river, but is constantly descending to
the water's edge, only to ascend again a few hundred yards
farther on.   It is in good repair, although the rocks are of
soft mud shales, and signs of landslips were frequent.   Wherever
possible maize is cultivated, but houses are few and far between.
The country strongly reminded me of that around Wênch'uan
Hsien in the upper Min Valley farther west.   Trees are very
scarce, the Wu-tung (*Sterculia*) being perhaps the most common.
The shrubs denote a dry (xerophytic) climate, nearly all having
small leaves, either thick or covered with a felt of hairs.   Of
these shrubs, *Abelia parvifolia*, *Lonicera pileata*, *Ligustrum
strongylophyllum*, and various kinds of Spiræa, are common.
Bushes of the wild Tea Rose are not infrequent.   Five li before
reaching Kai-ping-tsen, our destination for the day, we crossed
a clear-water tributary by a remarkably well-built stone-arch
bridge.   During the day we passed several " rope " bridges,

made of a single thick cable of plaited bamboo culms—sure signs of difficult borderland country. Near Shihch'uan Hsien we passed a bamboo suspension bridge, similar to the one already described ; at Kai-ping-tsen there is another such bridge. There was a fair amount of traffic on the road. Potash salts (lye), shingles, and oil-cake were the principal loads encountered, all being carried on men's backs, the first-named being the most common.

Kai-ping-tsen, alt. 3200 feet, is a small village of about fifty houses, situated on the left bank of a stream some 50 li north of Shihch'uan Hsien. A new, empty house afforded us comfortable lodgings ; the people were courteous, and made our brief stay with them very pleasant. A remarkably fine headstone, recently erected over the tomb of a much-respected widow, was the chief thing of interest in the village.

On leaving Kai-ping-tsen we continued to ascend the left bank of the stream through country similar to that of the previous day, for 30 li to the market village of Hsao-pa-ti. This village, all things considered, is of considerable size (about one hundred houses), with many farmhouses scattered around. The mountains are less rugged and steep, and are given over to the cultivation of maize. The houses are low, built of mud shales and roofed with slabs of slate. Market was in progress ; food-stuffs, fuel, and potash salts being the principal goods on sale. A bamboo suspension bridge spans the river and a road leads across country, ultimately joining with the main road between Shihch'uan Hsien and Mao Chou. On leaving Hsao-pa-ti the road deserts the river and ascends through maize fields over a rather low ridge. It then descends to a small tributary, after crossing which a steep climb of 1000 feet leads to the summit of another ridge. From this point we sighted the main stream again, flowing through a smiling valley, at the head of which nestles the village of Pien-kou, which was our destination for the day. This village proved a good 20 li from the ridge, though it looked close at hand. The road led through fields of maize to the valley, and finally across the river by an old, very shaky bamboo suspension bridge, which swayed tremendously and was really unsafe.

Pien-kou (Yüan-kou of the maps), alt. 3800 feet, is a market village of some importance, but a fire had recently destroyed half the houses. We had some difficulty in obtaining lodgings, the only decent place being full and the occupants unwilling to move. After a little time persuasive insistence won, and we settled down comfortably, if crowded. One of the occupants was down with fever. I dosed him with quinine, and supplied him with enough to last several days, much to his appreciation. This act got noised abroad, with the result that applications for medicine quickly became too numerous. Quinine is a drug much appreciated by Chinese, being about the only foreign medicine they have real confidence in.

The day's journey was said to be 70 li. It was long and uninteresting. The flora is miserably poor ; Alder being the only really common tree.

The road we were following ultimately joined the Mao Chou-Sungpan main road near Chên-ping kuan, about 160 li below the town of Sungpan. We could get no tidings of a road crossing to the Lungan-Sungpan highway, but all the same we felt sure of finding one. Thus far the route indicated on my map was all wrong, and we were left very much in the dark as to our actual whereabouts. However, I was long since accustomed to this state of affairs.

Leaving Pien-kou, a journey of 40 li brought us to Peh-yang ch'ang, a village of a dozen scattered, dilapidated houses. The road was distinctly bad in places owing to landslips. The rocks are mainly mud shales standing on edge. We followed the right bank of the river we had pursued from Shihch'uan Hsien for the first 22 li, then crossed over to the left bank by means of a shaky improvised bridge of two tree logs, the bamboo suspension bridge which formerly crossed the stream hereabouts having broken down. At this cross-over point resides a Chinese official, locally styled a Tu-ssu. This official was most courteous, helping us with advice and guidance to cross the stream.

The journey generally was a repetition of the two former days, through a rocky but uninteresting gorge. Wherever possible, maize is cultivated, and we noted two odd patches of rice. Houses are few and far between, and we met only a few coolies laden with potash salts, charcoal, and shingles.

VIEW FROM HOSTEL, LAO-TANG-FANG

The flora was not interesting, Alder, Pterocarya, and *Cornus controversa* are the only common trees. *Buddleia Davidii* is abundant by the stream side, and was in full flower. The Tea Rose also is fairly common. A Lily without bulbils, otherwise very like *Lilium Sargentiæ*, is plentiful in places. At Peh-yang ch'ang, alt. 4100 feet, we found a road leading off to the right, and connecting with the Lungan-Sungpan highway at Shui-ching-pu ; this we decided to follow.

Above our lodgings at Peh-yang ch'ang the river bifurcates, one branch, a clear-water stream, being locally adjudged the larger. It is up this stream the road connecting with the Mao Chou-Sungpan highway ascends. The people told us that this road was similar in character to the one we had followed thus far, but more difficult, especially since the proper bridges had nearly all been recently destroyed by floods. The cross-over to the Min Valley is near a place called Hwa-tsze-ling, where fine forests of Silver Fir and Spruce occur. Pien-kou is a considerable wine market, much of the product finding its way to Sungpan over this rough cross-country road.

A fatiguing march marked our first day's journey towards the Lungan-Sungpan highway. We made two long ascents and descents, and commenced a third ascent, putting up for the night at Hsao-kou, after covering 55 li. The second ascent was fully 2000 feet, and very steep, through maize fields, culminating in abandoned herb-clad areas. The descent was mainly through coppice and brush. Houses occur scattered here and there, wherever cultivation is practicable, maize being the staple crop ; the Irish potato and peas are also grown. The road proved difficult, but I had traversed worse.

The forests have been destroyed, brushwood now covering the uncultivated areas. Topping the loftier crags, and in inaccessible places generally, a sprinkling of conifer trees still exist, but we did not get near them. The vegetation generally is that common to the 5000 to 6000 feet belt in west Szechuan, but is less varied than in many parts I have visited. In the valleys Alder was common, and on the slopes the Varnish tree (*Rhus verniciflua*) and Walnut (*Juglans regia*) occur in quantity. In coppices the Davidia, both the hairy and glabrous varieties, is plentiful, but no large trees were noted. Throughout the bottom-lands and abandoned cultivated areas "Summer Lilac" (*Buddleia Davidii*) was a wonderful sight—thousands of bushes, each one with masses of violet-purple flowers, delighting the eye on all sides, the variety *magnifica*, with its reflexed petals and intense coloured flowers, being most in evidence. I gathered also an *albino* form, one small solitary bush, the only one I have ever met with. Forming a much-branched bush 4 to 8 feet tall, with rose-purple flowers, *Hydrangea villosa* was, next to the Buddleia, the most strikingly ornamental shrub. On moist rocky slopes plants of *Rodgersia æsculifolia* occur in millions. It was in the fruiting stage, but when in flower the acres of snow-white panicles must have presented a bewitching sight. Nowhere else have I seen this plant so abundant or luxuriant. The slender arching plumes of white flowers, produced by *Spiræa Aruncus*, covered acres of ground ; an apetalous Astilbe (*A. rivularis*) was also abundant, and worthy of note.

The hamlet of Hsao-kou, alt. 5900 feet, consists of three scattered houses, surrounded by maize plats, with remains of other ruined houses near by. It is encompassed on all sides by steep mountains, some of them culminating in lofty limestone crags and rugged razor-like ridges with pinnacled peaks—all of them inaccessible. At the back of the inn are a few Larch trees, and near by several large trees of a flat-leaved Spruce. The Hou-p'o (*Magnolia officinalis*) is cultivated hereabouts, and also around all of the houses we passed during the day. The innkeeper likewise cultivates a medicinal Aconite (*Aconitum Wilsonii*), which is valued as a drug in Chinese pharmacy.

We encountered only three men carrying goods during the whole day ; two were laden with potash salts, the third with the bark of a Linden (*Tilia*), used locally for making sandles. Evidences of forest fires were all too frequent during the day's march.

The next day rain ruined what otherwise would have been a more than ordinarily interesting march. From 7 a.m. until 2 p.m. we struggled up some 4000 odd feet to the summit of the pass leading across the Tu-ti-liang shan ; then descended another 4000 feet to the hamlet of Hsueh-po, where we secured lodgings in a large and good house. Rain commenced shortly before 11 a.m., and continued the rest of the day. Our perspective was limited to a few hundred feet ; now and again a strong gust of wind would scatter the mists, admitting momentary glimpses of cliffs and inaccessible peaks clothed with jungle and with occasional Conifer trees, but such views were rare.

The hamlet of Hsao-kou is very scattered, and we passed two or three more houses soon after leaving our lodgings. But after about 3 li houses and cultivation vanished, as did also the Buddleia and Hydrangea previously so abundant. The ascent, at first gradual, soon becomes precipitous, through a jungle growth of shrubs and coarse herbs. The latter with the thin brushwood is cut periodically and burnt. The ashes so obtained are placed in wooden vats fitted with sieve bottoms, boiling water is poured over them, and the liquid drains into tubs, where it is evaporated and salts of potash (lye) left as a residue. This product is packed in flasks and carried to market towns for sale. We passed several rude huts where men were engaged in this occupation. The road ascends a small torrent and is nowhere easy. By throwing logs across the stream and boggy places, lumber-men have succeeded in making some sort of a path. But crossing these wet, slippery logs was difficult. At one such crossing I slipped, but by jumping into the rock-strewn torrent somehow managed to avoid a nasty accident. Near the summit, and for some distance down the Lungan side of the pass, are split pieces of wood, arranged to form a long flight of shallow steps that assist the roadway materially. The descent after a few hundred feet becomes gradual, leading through open, park-like slopes, quite unlike anything I have encountered elsewhere in China. Now largely denuded of trees these glades are covered with grass, and horses, goats, and pigs are raised here in some quantity.

Formerly this range of mountains must have been covered with conifers, but the lumber-man's hands have been heavily laid on these forests. We passed none but small, decrepit specimens of no value. Hemlock, Spruce, and Silver Fir are all represented. The outstanding feature of the march was

the abundance of Cercidiphyllum trees.   Throughout the moist
slopes and park-like areas on both sides of the range this tree
is common.   Stumps of decaying giants abound, one of these,
which I photographed, measured 55 feet in girth!   This
specimen had been broken off some 30 feet above the ground,
and was a mere hollow shell, but still supported many twiggy,
leafy branches.   These stumps are relics of the largest broad-
leaved trees I have seen anywhere in China.   Growing interspersed with these remains were many specimens of the same
tree, 60 to 80 feet tall, 8 to 10 feet in girth, perfect in
outline, with myriads of neat, nearly round, bright green
leaves.   One of these was in young fruit, and for the first
time in my travels I secured specimens of the fruit of this
beautiful and interesting tree.   (Later I collected ripe seeds,
and this tree is now growing in the Arnold Arboretum, where
it promises to be quite hardy.   It proved to be a variety
distinct from the Japanese species.)

This tree (*Cercidiphyllum japonicum*, var. *sinense*) attains
to greater size than any other broad-leaved tree known from
the temperate zone of eastern Asia.   In size it is only
approached by its close ally, Tetracentron, which is also
common in the woods on the Tu-ti-liang shan.   A local
name for the Cercidiphyllum is " Peh-k'o," a name strictly
applied all over China to the Maidenhair tree (*Ginkgo biloba*).

The summit of the range is composed of mud shales, which
seem favourable to the growth of vegetation generally.   Between 8000 feet altitude and the summit *Rhododendron
calophytum* is extraordinarily abundant, trees 40 to 50 feet tall
and 5 to 7 feet in girth, with handsome cinnamon-brown bark,
cover many acres.   *Euptelea pleiosperma* and *Pterocarya hupehensis* are other interesting trees plentiful hereabouts.   The
bark of the last-mentioned tree is used locally for roofing
purposes.   Willows in many species are common ; the bark
of certain of these and also that of Linden trees is used by
the peasants for making sandals.   *Viburnum erubescens*, var.
Prattii, with pendulous panicles of white fragrant flowers,
followed by fruit which is at first scarlet and then changes
to black, is perhaps the commonest shrub.   Various Araliads,
Sorbus, etc., grow epiphytically on all the larger trees that

have a rough humus collecting bark. Maples in variety, Micromeles laden with fruit, and many other interesting trees were striking constituents of these woods. Tall growing herbs made a grand display, especially the apetalous *Astilbe rivularis*, *Spiræa Aruncus*, *Anemone vitifolia* with white and pink flowers like the Japanese Anemone, *Artemisia lactiflora* with large panicles of milk-white, fragrant flowers, Balsams (*Impatiens*) with yellow, pink, and purple flowers ; mixed with them also were Meadow Rue (*Thalictrum*), Aconites, many Senecios, and *Meconopsis chelidonifolia* growing about 3 feet tall with clear yellow flowers, saucer-shaped and 2½ inches across. Acres of the country-side are covered by these various herbs.

There was indeed plenty to interest one ; the flora of this region is undoubtedly rich, and it was most unfortunate that the rain prevented an exhaustive investigation.

Hsueh-po, alt. 6000 feet, consists of a few houses surrounded by high mountains with a good-sized torrent, which rises near the head of the pass, and flows through the narrow valley. Maize is cultivated as the staple crop. The Hydrangea and Buddleia previously noted ascend to this altitude, and were a wealth of blossoms. Alder also extends to this point ; Poplar likewise. This latter tree has a very graceful port and the leaves have red petioles and veins when young.

Our lodgings were good and weather-proof, which was fortunate, since it rained heavily the night through, and until eleven o'clock in the forenoon of the next day. Afterwards it was fair, but threatening, heavy clouds and mists obscured the country-side from our view. Around the inn are several trees of a handsome, flat-leaved Spruce (*Picea ascendens*) with pendulous branchlets. This tree, known locally as " Mê-tiao sha or sung," is the most esteemed timber tree in these parts. The trees are felled, hewn into planks about 25 feet long, 5 inches thick, and 12 inches broad, and carried on men's backs to a point on the river whence it is possible to float down rafts. Lumbering is a very considerable industry in these mountains, the timber finding its way to Chungpa. This fine Spruce was fruiting freely. (Later I secured plenty of seed, and successfully introduced it into Western gardens.)

On leaving Hsueh-po, we crossed the torrent and descended

the left bank.   At K'ung chiao the torrent is joined by another of equal size, the united waters forming a fine clear-water stream.   From this point downwards rice is cultivated.   The stream continues to receive affluents, a very considerable one joining it at Tu-tien-tsze.   At Peh-mu chiao, 10 li above Tu-tien-tsze, the timber logged in the surrounding mountains is made into rafts and floated down.   Just below Shui-ching-pu the stream unites with the main branch of the Lungan River (the Fou Ho), and the rafts are floated down past the city of Lungan to Chungpa, a large village of vast commercial importance, in direct water communication with distant Chungking, it being within the Kialing River system.

Tu-tien-tsze is a small market village and a Roman Catholic Mission centre.   This Church has a strong following throughout the region we had traversed from An Hsien.   The country folk everywhere in this part were most courteous and civil.   This, I think, is probably due to the influence wielded by the self-sacrificing priests of the Roman Faith.   But whatever the cause, I shall always retain pleasant memories of the people encountered everywhere in this little-known region.

The road proved easy all day, usually skirting the mountains well above the stream.   At Tu-tien-tsze a cross-country road leads to Lungan Fu, some 130 li distant.   Ten li below Tu-tien-tsze we crossed to the left bank of the stream by a covered bridge.   Descending a few miles and crossing a promontory we reached the main river (Fou Ho) opposite Shui-ching-pu.   Ferrying across to this village, we found lodgings in a large house owned by a Shensi man of the Mohammedan persuasion.

Shui-ching-pu, alt. 4200 feet, is a market village of about 200 houses, situated on an alluvial flat, surrounded by mountains largely under cultivation.   A river of considerable size, which brings down an unusual quantity of detritus, joins the main stream on the left bank immediately below the village.   A road ascends this stream, leading to Wên Hsien in Kansu province.   It was said to be difficult, traversing a mountainous region peopled with Sifan.   Iron is a local product of some importance hereabouts.   Gold is also mined in the neighbourhood.   The quartz, after it has been broken into small pieces, is pounded into dust in mills like those commonly used for hulling rice. The dust is washed and the gold separated by means of quicksilver. Placer mining is carried out all along this Lungan River by unemployed peasants, but the yield is small. In 1904, when I first journeyed to Sungpan by way of Mien Chou, Chungpa, and Lungan, the officials were endeavouring to put a stop to placer mining. Placards were posted forbidding the people to wash for gold, on the ground that landslips were caused through the removal of the rocks, etc., on the foreshore.

From Hsueh-po to Shui-ching-pu is said to be 60 li. The valley which we traversed is all under cultivation; farmsteads are general after Peh-mu chiao is reached. Alder, Walnut, and Poplar are the common trees, with Pear, Plum, and Peach trees around houses. In a garden I saw one magnificent specimen of the Crêpe Myrtle (*Lagerstrœmia indica*), 25 feet tall, 2½ feet girth, just one luxuriant mass of carmine-red flowers. Here and there the moist rocks are beautifully carpeted with ferns, *Woodwardia radicans, Blechnum eburneum*, and Maidenhair being particularly rampant. The Buddleia and Hydrangea, previously mentioned, are abundantly present, and were a wealth of pleasing flowers.

At Shui-ching-pu we joined the highway between Lungan Fu and Sungpan. The intrepid Captain W. J. Gill,[1] in June 1877, was the first Occidental to traverse this route. Since that date several travellers and missionaries have been over this road, but the total is small.

My first journey over this highway was, as mentioned above, in 1904. At that time I had no camera, and the recollection of the wonderful scenery had much to do with my second journey to these parts in 1910. I saw the country through the eyes of a botanist, and for this reason I hope a continuance of this narrative will prove justifiable.

Leaving Shui-ching-pu about 7 a.m., we saunteringly covered the 50 li to Hsao-ho-ying by 4 p.m. The road ascends the left bank of the stream for some 20 odd li to a point just above the small village of Yeh-tang. At this place the river is joined by another of nearly equal size on its right bank. A by-road ascends this tributary and leads across the mountains

[1] *River of Golden Sand.*

through country sparsely peopled with Sifans, and connects with the Mao Chou—Sungpan highway a few miles below Sungpan Ting. The road we followed crosses the left affluent of the Fou River by means of an iron suspension bridge 24 yards long, erected immediately above the union of the two streams. A few li beyond this place the road plunges into a wild gorge. The scenery is wonderful. Limestone cliffs clad with vegetation rear themselves 1000 to 2000 feet above the torrent which hereabouts rushes headlong over huge rocks. Wherever possible, maize is cultivated on the slopes and rice in the bottom-lands. We crossed to the right bank by a covered wooden bridge just below a place where landslips have produced a series of cataracts. About 3 li below Hsao-ho-ying the gorge suddenly opens out, leaving room for a small circular valley, in the middle of which the walled village above named is situated. Viewed from this point where there is an old gateway, the village presents a charming picture of peace and plenty locked in by precipitous mountains. On entering the village, however, one is quickly disillusioned. Abject poverty is only too apparent. The one main street is broad, flanked by more or less ruined houses, with much of the land within the walls given over to maize plats. The people are in keeping with their dilapidated surroundings.

Hsao-ho-ying, alt. 5300 feet, signifies " Camp on the Small River." It is an ancient garrison village. Eighty years ago some 700 soldiers were quartered here. This number was speedily reduced as the surrounding country was conquered. To-day the garrison is put down at 40 men, but it is doubtful if even this number remains. Three yamêns belonging to military officials of low rank are the only respectable buildings in the place.

At Shui-ching-pu we were assured we could exchange silver at Hsao-ho-ying. This proved a fable and landed us in an awkward dilemma. However, " Mo-li-to " (*Fata viam invenient*), as the locals have it !

The flora of the day's journey was not particularly rich, though we passed many plants of interest. Around Hsao-ho-ying, the Walnut (*Juglans regia*), Varnish (*Rhus verniciflua*), Poplar, Apple, Pear, Plum, Peach, and Tu-chung (*Eucommia ulmoides*) are commonly cultivated. By the side of the torrent the Buddleia was again a wonderful sight. In a temple yard near Yeh-tang is a magnificent tree of *Meliosma Beaniana*, about 60 feet tall and 12 feet in girth, the head being fully 80 feet in diameter. The pinnate leaves produce abundant shade. This tree was covered with small pea-like purple fruits which later afforded me a supply of ripe seeds. (The pinnate-leaved members of this small family are all handsome trees, and none was in cultivation previous to my explorations. I have succeeded in introducing three species, all of them promising to thrive under cultivation. One, *M. Veitchiorum*, is now flourishing just within the main entrance to Kew Gardens.)

From Hsao-ho-ying to Shuh-chia-pu, 30 li, the road ascends a narrow valley which is without special interest, the bottom lands and lower slopes being cultivated with maize and buckwheat. Houses occur at intervals. Just above Shuh-chia-pu, a poverty-stricken hamlet of about a score of houses, the river bifurcates. The road ascends the left and larger branch, plunging immediately into a narrow gorge. The track, all things considered, is good, though there is room for improvement. The scenery in this gorge, for magnificent, savage grandeur, would be hard to surpass. The cliffs, chiefly limestone, are mostly sheer, and 2000 to 3000 feet high. Wherever vegetation can find a foothold it is rampant, and a luxuriant jungle of shrubs clothes all but the most vertical walls of rock. By the side of the torrent coarse herbs, shrubs, and small trees abound. The mountain crests and ridges are covered with Spruce and Pine. Now and again glimpses of vicious-looking, desolate peaks, towering above the tree-line, were obtainable. The waters of the torrent roar and dash themselves into foam in their passionate endeavour to escape to more open country. In more peaceful stretches the river describes a series of S-curves with shingly areas covered with *Myricaria germanica* and *Hippophaë salicifolia* (Sallowthorn), jutting out into the current. In one place the cliffs recede somewhat, leaving room for a narrow valley, where three or four peasants' huts are pitched. Around these cabins forlorn patches of maize, buckwheat, cabbage, Rhubarb (*Rheum palmatum*, var. *tanguticum*), and Tang-kuei (*Angelica polymorpha*, var. *sinensis*) are cultivated. The

abandoned clearings are covered with coarse herbs, among which
Senecio clivorum, growing 4 to 5 feet tall, with its golden yellow
flowers, was prominent. *Astilbe Davidii* also abounds ; likewise the Buddleia. A sub-shrubby Elder, growing 3 to 5 feet
tall, with masses of salmon-red fruits, was a pretty sight in
all the more open moist places. (The species proved to be
new and has been named *Sambucus Schweriniana*, Rehder, in
Plantæ Wilsonianæ, Part II. p. 306 (1912).) The vegetation
indeed is rich and varied, and a large harvest of specimens
rewarded the day's labours. After scrambling some 30 li
along this gorge we reached the hostel of Lao-tang-fang just
as night was closing in. We encountered considerable traffic
on the road. Coming from Sungpan were coolies laden with
medicines, sheep-skins, and wool. Journeying thither the coolies
were laden chiefly with wine in specially constructed tubs,
preserved pork and rice. Lao-tang-fang, alt. 7600 feet, consists
of one large new hostel, not quite completed at the time of our
visit ; a long row of " bunks " are built along one side, with
benches for the accommodation of loads on the other. The
whole structure is of wood, the roofing being of shingles badly
laid. The mud floors were very damp, and vegetation was
springing up in the corners and under the bunks. Skins of
Serow and Budorcas served as mattress on the bunks, or
settees, and no two of these skins exhibited the same coloration.
Both animals are said to be common in the neighbourhood,
more especially the Serow. The Parti-coloured Bear, or Giant
Panda, also occurs here in the Bamboo jungles.

The hostel was full to overflowing and undoubtedly supplies
a much-needed want. For the sake of future travellers, if for
no other reason, I heartily hope success attends the landlord's
venture. Formerly a most miserable structure occupied this
site, and I have unpleasant memories of a night spent there in
1904. Except for a tiny cabbage-patch there was no sign
of cultivation around the hostel, but clearings were being
made for the purpose of cultivating Tang-kuei and other
medicines. The view from this spot is savage and grand
beyond power of words. On all sides are precipitous mountains, towering 3000 feet and more above the torrent, all more
or less densely forested. Almost facing the river is a limestone

HSUEH-PO-TING, WITH SNOWS; BED OF STREAM ENCRUSTED WITH LIME DEPOSITS

cliff with upturned strata on edge, sheer and bare of vegetation. Behind this is another nearly vertical slope covered with stark, dead Conifer trees. In the distance, looking back on the road we had followed, bare, vicious-looking peaks, probably 14,000 to 16,000 feet high, were visible. All around the hostel the lesser slopes are covered with impenetrable forest of broad-leaved deciduous trees. The higher parts and the crags are clothed with Conifers, tall, slightly branched trees of no great size—altogether a wonderful scene of natural beauty, at present undefiled by the hand of man.

It was cold during the night ; the wind played freely through the unfinished structure, and the thickest of clothing was needed in order to keep warm.

The next day we made a later start than usual, and travelling most leisurely covered the 40 li to San-tsze-yeh before 5 p.m. The journey was one long scramble through a continuation of the savage ravine. The chairs had to be carried piecemeal, and all of us reached our destination very much fatigued. We enjoyed a gloriously fine, sunny day, the narrow streak of sky visible from the bed of the ravine being of the purest Thibetan-blue. The camera was kept busy and I secured a fine set of views, but so steep is the country and so dense the jungle that it was impossible to photograph trees.

The rock-strewn torrent, with its thundering, seething waters, occupies practically the entire bed of the ravine, leaving scant room for the road which winds along its banks. We crossed this torrent many times, either by fording it or by means of half-rotten log bridges. Luckily the waters were low and caused us no trouble. In 1904 I ascended this ravine shortly after heavy rains, and have the liveliest recollections of the difficulties encountered. Much of the road and many of the bridges had been washed away, making it necessary to hew a pathway through the jungle and improvise bridges by felling trees in several places.

No words of mine can adequately depict the savage, awe-inspiring scenery of this wild ravine. Stupendous limestone cliffs, 3000 to 4000 feet high, often too steep for the scantiest vegetation to find a foothold, but more generally sparsely or plentifully forested, wall in the torrent and its accompanying

roadway. Waterfalls abound, but lateral torrents are few.
The flora is very rich, but largely inaccessible. Practically
all the trees, shrubs, and herbs common to the 7000 to 9000
feet belt occur here. Conifers are the principal trees. Silver
Fir, Spruce, Hemlock, Larch, White Pine, Juniper, and Yew
are all represented. The Pine (*P. Armandi*) is the commonest
tree up to 8500 feet, clinging to the sheer cliffs in a remarkable manner. With its stunted branches and short leaves it
was hardly recognizable, suggesting a green Maypole rather
than a Pine tree! Many of the Spruce and Silver Fir were
fruiting freely, the erect, violet-coloured symmetrical cones of
the latter being very handsome. Larch (*L. Potaninii*) abounds,
overtopping all the other Conifers, but the trees are small. All
the Conifers are hereabouts designated " Sung-shu " (liter-
ally Pine trees), but the timber of the Larch, flat-leaved Spruce,
and White Pine, valued in the order given, are most prized for
building purposes generally. Of the broad-leaved deciduous
trees, Maple (*Acer*), Linden (*Tilia*), and Birch (*Betula*) are the
most common. A few Poplar occur, but Oak is exceedingly
rare, the few noted being scrubby evergreens of no great
beauty. The variety of shrubs is very great, all the more
woodland genera being rich in species. Sorbaria, with its large
panicles of white flowers, was one of the most attractive.
Spiræa, Viburnum Lonicera, Rubus, Philadelphus, Sorbus, and
many other families, made a fine display either with their
flowers or fruit. Strong-growing herbs, like the various species
of Senecio, Astilbe, Aconitum, and Anemone, cover miles of the
roadside. In shady places the handsome Maidenhair fern,
Adiantum pedatum, was a charming picture ; in sunny spots
the lovely *Gentiana purpurata*, with intense carmine-red flowers,
was a sight never to be forgotten.

About 10 li below San-tsze-yeh the ravine widens out into a
narrow valley, with the mountain-slopes on the left bank of the
torrent less precipitous and grass-clad. We passed the ruins of
some old forts, and shortly afterwards a Sifan hamlet consisting of three or four farmsteads, with numerous prayer-flags
fixed on the roofs. In the tiny valley wheat, barley, buckwheat, oats, peas, and broadbeans are cultivated, and the
crops were ready for harvesting.

San-tsze-yeh, alt. 9200 feet, consists of ruinous hovels built on a level with the infant stream which at this point breaks up into three equal branches, all of which have their source in the near neighbourhood.   Looking back on the route we had traversed we saw that all the higher peaks are barren and desolate, the highest of all being flecked with snow.   The whole plexus is made up of the spurs and buttresses of the mighty snow-clad Hsueh-po-ting.   To the north-east from San-tsze-yeh are other tremendous peaks, bare, barren, and uninviting in appearance.   The aspect of the country around this hamlet is purely Thibetan.   The scant crops and abject poverty of the inhabitants speak plainly of a country where altitude and climate set agricultural skill and industry at defiance.   Such regions the Chinese abhor and cannot colonize.   The pastoral Sifan, with their herds of cattle and sheep, remain masters of the soil though politically subject to Chinese authority.   The conquest of this wild region must have been a most difficult task and speaks volumes for the military genius which accomplished it.

During the night at San-tsze-yeh I had a violent attack of ague, probably caused by a chill, which culminated in a fit of vomiting.   This seizure and the howling of many dogs were against a good night's sleep.   In consequence we took things very gently the next day, and I used my chair much more than usual.

Twenty-five li above San-tsze-yeh, to the right of the stream which descends the narrow valley, there is a most interesting place.   A torrent heavily surcharged with lime descends from the eternal snows of the Hsueh-po-ting, depositing along its course thick lime encrustations of creamy white.   The place is considered holy by the Sifan, to whom any natural phenomenon strongly appeals.   A temple has been erected here and a series of some fifty tarns constructed by leading the waters from the stream and making small semicircular dams.   All are at slightly different levels, and the waters as they flow from one to another continue to build up the dams by leaving deposits of lime behind.   The bed of each tarn is creamy white, but owing to the light being reflected in different colours, according to the varying depth of each, an attractive scene of many-coloured waters is presented.   Some are clear azure blue, others creamy white, pink, green, purple, and so on.   The temple is called " Wang Lung-ssu " (Temple of the Dragon Prince), and it is fitting that the Sifan, children of nature as they are, consider the place holy.   Near the temple the waters have built up a wonderful series of waterfalls, and every fallen tree and bush obstructing the waters is speedily encrusted with lime. Above the temple the stream is fully 80 yards wide, and the bed is creamy white with soft encrustations of lime, the ripple marks being beautifully defined.   These lime-deposits extend for a mile or two and present a most striking scene.

From the bed of this stream, a short distance above the temple, a fine view of the snow-clad Hsueh-po-ting is obtainable.   The face visible carries but little snow, and immediately below the glaciers are wonderful cliffs of red-coloured rock.   In contrast the colour-effects are most remarkable.   There was said to be another temple some few li higher up towards the snows, but I was too fatigued to visit it.

All around Wang Lung-ssu are fine forests of Spruce, Silver Fir, Birch, with miscellaneous trees and shrubs.   In the vicinity of the lime-deposits the trees look very unhealthy, many are bleached and dead, others yellow and dying.   From the vegetation it is evident that these lime-deposits are recent and spreading rapidly.   A few Rhododendrons occur on the margins of the stream and in the woods, but are not happy. Right by the water's edge I gathered *Arctous alpinus*, var. *ruber*, a tiny alpine shrub with red fruit closely allied to the Blueberries, and found also near the glaciers in British Columbia !   This pretty little plant, only some 4 to 6 inches high, is quite common hereabouts, but had not before been recorded from China. Near the tarns *Cypripedium luteum*, a yellow-flowered counterpart of the North American Moccasin flower (*C. spectabile*), is very abundant.   (Later I succeeded in introducing live roots of this species to the Arnold Arboretum, where plants are now growing.)

The forests of this immediate neighbourhood are rich in fine Spruce trees, 80 to 150 feet tall and 6 to 10 feet in girth, with short branches producing a spire-like effect, are characteristic of the region.   The Silver Fir are less noteworthy, but, like the Spruce, were fruiting freely. (Both were subsequently introduced to cultivation.) Larch overtops all other trees, reaching its limits at about 12,000 feet altitude. The vegetation of the ranges flanking the narrow valley, up which the main road ascends, presents a remarkable contrast. The range to the left of the stream, above 10,000 feet altitude, is covered only with scrub and grass; whereas the range on the right bank is heavily forested up to altitude 12,000 feet. Early in the afternoon, after covering 40 li, we reached the lonely hostel of San-chia-tsze, alt. 12,800 feet, situate some 600 feet below the head of the pass. During the first 25 li of the day's march we passed several large farmhouses, but nearly all are deserted and falling into ruins. Around these houses a few plats of wheat, barley, flax, and Irish potato are cultivated; also cabbage, garlic, and other vegetables in minute quantities. Tobacco (*Nicotiana rustica*), in small quantities for household use, is grown around San-tsze-yeh, and the crop looked very happy. These sporadic attempts at cultivation represent the vain and futile efforts of the Chinese settlers to eke out an existence from the inhospitable soil. This side of the pass is evidently much colder than the Sungpan side, since there, at greater elevations, good crops of wheat, barley, and peas can be raised.

Apart from the forests already mentioned, herbs dominate the flora. A great variety were still in flower, the various species of Senecio and Gentiana being most striking. *Gentiana detonsa*, a slender plant a foot and more tall, with numerous large deep blue flowers, looked particularly happy, flaunting its blossoms in the sun. On rocky screes the yellow-flowered *Clematis tangutica* is abundant and was covered with its top-shaped blossoms. The hedges bordering the fields are composed chiefly of Wild Gooseberry and *Sorbaria arborea*: the latter was in full flower. In copses by the stream, up to 11,500 feet, Hornbeam, Cherry, Red Birch, Willow, Maple, and Hazel-nut are common. The Hazel-nut is mainly *Corylus ferox*, var. *thibetica*, a variety having a spiny fruit closely resembling that of the Sweet Chestnut (*Castanea*).

The hostel of San-chia-tsze is maintained for the accommodation of travellers, and a posse of soldiers is stationed here to keep down banditti. The hostel is a roomy but miserable

cabin, built of shales and roofed with shingles held down by
stones. The floor is of mud and very uneven; there is no outlet
for smoke, save the doorway, and no windows. At midday
a candle was necessary to avoid falling over things when moving
about indoors. During different visits I have suffered many
days and nights in this lonely spot, on one occasion being
snowed in for three consecutive days. The cabin is situated
on a narrow sloping valley running nearly east and west, a
mile or so above the tree-limit, flanked on the northern side
by a ridge of stark, crumbling rocks. To the south the range
culminates in bare peaks and eternal snows of the Hsueh-po-
ting. The moorland country all around is typical of Eastern
Thibet, so perhaps a few details are permissible. The treeless
spurs and valleys are covered with extensive heaths of scrub,
made up of several species of Spiræa (including *S. mollifolia*,
S. alpina, and *S. myrtilloides*), *Sibiræa lævigata*, *Lonicera
hispida*, *L. chætocarpa*, *L. prostrata*, *L. thibetica*, and others,
several Barberries, Currants, shrubby Potentillas, Astrag-
alus, Sallowthorn, small-leaved, twiggy Rhododendrons, and
Juniper. As the altitude increases, one by one these shrubs
give out until only the Juniper is left. This ceases about 15,000
feet; alpine herbs ascend another 1000 feet, and the limit of
vegetation is roughly 16,000 feet. The Juniper scrub is from
1 to 2½ feet tall, very dense and difficult to traverse, but
furnishes excellent fuel. Mixed with this scrub are herbs
in great variety, the Poppyworts (*Meconopsis*) being particularly abundant. Possibly the commonest herb between
12,500 feet and 14,000 feet is *Meconopsis punicea*, a lovely
species having large, dark-scarlet nodding flowers. (It was
from near this vicinity that I succeeded in introducing this
plant in 1903.) The violet-blue flowered *M. Henrici* is common
between 13,000 feet to 14,000 feet, but much less so than around
Tachienlu. The prickly *M. racemosa*, with blue flowers, is
plentiful in rocky places between 13,000 feet to 14,500 feet.
From 11,500 feet to 13,000 feet the gorgeous *M. integrifolia*,
growing 3 feet tall, with its peony-like, clear yellow flowers
8 to 11 inches across, occurs, but is not plentiful. The intense
colours among alpine flowers everywhere is well known, and
this region is no exception. The yellow is mostly supplied by

LOOKING EAST-SOUTH-EAST FROM HSUEH SHAN PASS, RUINED FORT IN FOREGROUND

Senecio, Saussurea and other *Compositæ*, slender growing Saxifraga, etc. The blue and purple by various Aconites, Larkspurs, and Gentians; among the latter *Gentiana Veitchiorum*, with large erect flowers, covers large areas. The Lousewort (*Pedicularis*) and Fumewort (*Corydalis*) are represented by many species, having flowers embracing all the cardinal colours. Primulas occur, but not in many species. Androsace, Sedum, Cyananthus, and other alpine genera abound.

Large flocks of sheep are pastured on these uplands, but yak are not kept in quantity hereabouts. There is not much variety of game. Blue sheep are common, Budorcas are found near the timber line; on the higher crags occasional flocks of Goa, or Thibetan gazelle, occur. Snow-partridge, Thibetan Hazel-hen, Snow-cock and allied game-birds, together with Thibetan Hares, are fairly numerous. The Wolf is the only carnivorous animal really common.

The Hseuh-po-ting snows are visible on clear days from the wall of Chengtu city, and are accounted the "Luck of the Plain." The Chinese claim that so long as snow covers this peak the prosperity of Chengtu and its surrounding plain is assured. It was a perfect moonlight night on the occasion of my last sojourn at San-chia-tsze, and late in the evening I beheld the "Luck of Chengtu," with its crown of eternal snow lit by the radiant moonlight. The loneliness of the region, the intense stillness on all sides, and the wonderful peak with its snowy mantle, made a most impressive scene.

A glorious morning followed a perfect night. From the head of the pass (alt. 13,400 feet) I obtained further good views of the Hsueh-po-ting, bearing west-south-west and secured some photographs. The peak is probably 22,000 feet high, in shape an irregular tetrahedron, the south-west slopes carrying enormous snow-fields. The north-east face is very steep and carries but little snow. The surrounding peaks are bare and desolate looking; no vestige of life was discernible, and the whole scene was lonely, most forbidding, even awesome, though bathed in brilliant sunshine.

Below San-chia-tsze are the stone ruins of an old fort and stockade, relics of ancient warring times, but now covered with various herbs, especially Saxifrages, which were masses of yellow, and other coloured flowers. The head of the pass is marked by a ruined tower and fort, from the summit of which Thibetan prayer-flags waved. That robbers still haunt these regions was brought home to us by the sight of a partially covered coffin near the head of the pass. A few weeks before, a poor coolie, bound towards Lungan Fu to purchase rice, was attacked here, robbed, and killed. The bandits got clear away. The coolie's " pai-tzu " (a framework for carrying loads on) and various appurtenances lay on top of the coffin and remain to tell the story of the crime. All around are grassy areas, covered at the season of our visit with blue and yellow alpine flowers.

At the head of the pass small boulders of sandstone, marble, granite, and other rocks lay scattered around. Just below are beds, which resemble coal-ashes, probably of volcanic origin.

From the pass we dropped down into a valley which quickly led to fields of golden wheat and barley. The crops were ripening, and here and there the reapers were busy. Passing a ruined fort, several Sifan farmsteads, and a lamasery, the road led to the summit of a grassy ridge. Descending a few hundred feet we sighted the city of Sungpan nestling in a narrow, smiling valley, surrounded on all sides by fields of golden grain, with the infant Min, a clear, limpid stream, winding its way through in a series of graceful curves. In the fields the harvesters were busy, men, women, and children, mostly tribesfolk, in quaint costume, all pictures of rude health, laughing and singing at their work. Under a clear Thibetan-blue sky, the whole country bathed in warm sunlight, this busy scene of agricultural prosperity gladdened the hearts of all of us, fatigued and exhausted as we were from the hardships of our journey through savage mountains with their sublime scenery and wonderful flora.

CHAPTER XI

SUNGPAN TING

THE LAND OF THE SIFAN

THE city of Sungpan is situated on the extreme north-west corner of Szechuan, about long. 103° 21′ E., lat. 32° 41′ N., at an altitude of 9200 feet, and is the farther-most outpost of Chinese civilization in this direction. The surrounding country, more especially to south-west, west, and north-west, is inhabited by Sifan, a people concerning which very little is known. Originally established as a military post after the conquest of the neighbouring regions by the Emperor Kienlung about A.D. 1775, Sungpan has developed into a most important trade entrepôt. It is a city of the second class (styled " Ting "), but the head civil official has the local rank of prefect, his full title being " Fu-I-Li Min-Fu," which signifies " the Bar-barian-cherishing, Chinese-governing Prefect." This fanciful title has reference to the official's control over the neighbouring Sifan tribes—a control which is purely nominal. The military importance of this stronghold is still fully recognized, and its strategic value is beyond question. A Chinese general (Chen-tai), in command of ten regiments, has his headquarters here, with jurisdiction extending south to Kuan Hsien, east to Lungan Fu, and north-east to Nanping in Kansu province.

The town is most picturesquely situated, occupying con-siderable space in a narrow, highly cultivated valley flanked by steep mountain-slopes 1000 to 1500 feet high. The Min (Fu) River, which takes its rise some 35 miles to the north, winds a circuitous course down the valley and flows through the town in an S-curve, entering and leaving through the city walls at unfordable points. On the western side the town is backed by a steep slope, up two sides of which a wall is carried.

The west gate of the city is situated at the top of this slope, and is exactly 1000 feet above the river. Save for a yamên and a temple or two the whole of the mountain-slope within the walls is given over to terraced cultivation, the city proper being clustered in the valley alongside the river. The wall surrounding three sides of the city is very substantially built of brick, being fully 20 feet thick and more high, but that which ascends the mountain-sides is in places only 2 feet thick and 4 feet high ; a steep ravine, however, immediately outside this wall, affords additional protection. Since the Chinese first established themselves here the town has undergone many vicissitudes. Time and again the Sifan have swept down upon it, captured it, and massacred all who fell into their hands. So frequent have been these attacks, and so great is the Chinese dread of treachery on the part of the Sifan, that it is only within the last few years that any of these people have been allowed to remain overnight within the city walls.

In 1910 Sungpan had a resident civil population of about 3000 people, and a floating population equalling, if not exceeding, this number. The houses are nearly all of wood, generally well built, with rather curiously-carved porticoes ; the timber employed for building is mostly Juniper, which is floated down the Min River from a point some 15 miles to the north-north-east. In October 1901 the city was two-thirds destroyed by fire, but on the occasion of my last visit in 1910 the devastated area had been practically rebuilt. The streets are badly paved, ill-kept, and the city possesses no buildings of architectural beauty. Near the south gate the military section of the town is situated, and a considerable amount of market-gardening is carried on there. The people are very fond of flowers, nearly every house boasting some in pots, on the walls, or in borders. Stately Hollyhocks, with multi-coloured flowers, are a feature. With these are generally planted Tiger Lilies, Chinese Asters, and small-flowered Poppies, the whole making a bright and pleasing effect. The Chinese Aster (*Callistephus hortensis*) is wild in the neighbourhood ; the Poppy is a species closely allied to *Papaver alpinum*. The population is mainly Mohammedan Chinese, who carry on a remunerative barter-trade with the surrounding

THE CITY OF SUNGPAN TING

tribes.   Tea is the all-important medium employed, this commodity and a few odd sundries being taken in exchange by the tribesmen for their medicines, skins, wool, musk, etc. During the month of July a fair is held annually for trade purposes.   The people from far and near attend this fair, a vast amount of business being transacted.   Trading caravans also make long journeys into the country north-west to the borders of the Kokonor regions.   Wool, sheep-skins, and various medicines in great quantity are exported from Sungpan to different parts of China.

The trade passing through Sungpan is, I am convinced, not only greater than has been estimated, but is increasing annually.   In 1903, on the occasion of my first visit to this town, I enjoyed the companionship of W. C. Haines-Watson, Esq., then Commissioner of the Imperial Maritime Customs at Chungking.   This gentleman investigated the trade of this region, estimating the exports to Thibet at Tls. 801,000, and those into China at Tls. 512,000 (" Journey to Sungp'an," *Jour. China Branch Roy. Asiat. Soc.*, 1905, xxxvi).   Our visit occurred before the city had recovered from the disastrous fire of 1901, and trade was suffering in consequence.   In 1910 trade was evidently booming.   I have no figures to guide me, but comparing the two visits I would put the trade with China alone at a million taels.   This trade has three outlets : one, east, via Lungan Fu to Chungpa ; another, south-east, via Mao Chou, Shihch'uan Hsien to An Hsien ; the third through Kuan Hsien to Chengtu.   The first two routes afford water communication from Chungpa and An Hsien respectively, with Chungking on the Yangtsze River.   By these routes most of the goods intended for Chungking and beyond are conveyed.   The trade via Kuan Hsien is mainly with Chengtu and other cities on the plain.   This latter trade route has been looked upon as the most important, whereas it is really less so than either of the other two outlets.

The late Captain W. J. Gill in 1877 was the first Occidental to visit Sungpan.   Since that date several foreigners have paid visits, and missionaries of Protestant denominations have made abortive attempts to establish stations there.   I have visited this place three times, and on each occasion enjoyed the stay and departed with regret.   Did the Fates ordain that I should live in Western China I would ask for nothing better than to be domiciled in Sungpan.   Though the altitude is considerable the climate is perfect, mild at all times, with, as a general rule, clear skies of Thibetan-blue.   During the summer one can always sleep under a blanket, in winter a fire and extra clothing are all that is necessary.   Excellent beef, mutton, milk, and butter are always obtainable at very cheap rates. The wheaten flour makes very fair bread, and in season there is a variety of game.   Good vegetables are produced, such as Irish potatoes, peas, cabbages, turnips, and carrots, and such fruits as peaches, pears, plums, apricots, apples, and Wild Raspberries (*Rubus xanthocarpus*).   Nowhere else in interior China can an Occidental fare better than at Sungpan Ting. With good riding and shooting, an interesting, bizarre people to study, to say nothing of the flora, this town possesses attractions in advance of all the other towns of Western China.

The valley, which varies from ¼ to ½ mile in width, and the mountain-slopes, rising from 1000 to 1500 feet above, are given over to wheat and barley cultivation, with occasional fields of peas and flax, the latter being valued for its seeds, which yield an oil used as an illuminant.   In the latter half of August the whole country-side is one vast sheet of golden grain bending to the wind.   This grain is reaped, leaving a generous stubble, which is immediately ploughed under.   The ploughs are simple, consisting of an iron-shod shear, a straight handle of wood, and a long shaft, to which is harnessed a couple of oxen or half-bred yak.

In harvesting the grain, tribesfolk (chiefly Po-lau-tzu), who come from the upper reaches of the Tachin Ho, many days' journey to the west-south-west, play an all-important part.   Every year these people visit this region for the express purpose of this work, and are, in fact, indispensable.   As the crop is reaped it is tied into little sheaves and stacked ears downwards on high hurdle-like frameworks (Kai-kos) to await threshing.   The threshing is done by wooden flails, both men and women taking part in the work.   The corn is ground in mills driven by water-power.

The name " Sungpan " has reference to forests of Spruce

TWO SIFAN

and Fir and the circuitous course of the river Min (Fu). The river still pursues its winding course, but the forests have long since disappeared. It is only in temple grounds and among tombs that any trees remain. The mountains are absolutely treeless, where not under cultivation they are covered with scrub and long grass. The outer crust of the mountains consists of a rich flaky loam, probably of glacial origin, rather heavy, but specially adapted for cereal cultivation. In the grass and scrub Pheasants are very plentiful in the neighbourhood of cultivation, so also is a long-eared, light-grey-coloured Hare. Musk Deer, Wapiti, and White Deer occur in the neighbourhood. On the moorlands a Marmot, called " Hsueh-chu " (Snow-pig), burrows in large colonies.

North-west of Sungpan is the Amdo country, a region of grasslands. The Chinese designate it " Tsaoti," which may be interpreted " Prairie." This region is made up of rolling country above 11,000 feet altitude, where vast herds of cattle, sheep, and many ponies are reared. A great part of this region is peopled with pastoral Sifan, but the more remote parts are in the hands of nomads belonging to Ngo-lok and Nga-ba tribes, of evil reputation as robbers and bandits, dreaded alike by Chinese and the more peaceably inclined Sifan. These robber tribes are of Tangut origin, having their headquarters around the Kokonor region. Being of nomadic habit they wander far afield, and rob caravans and kill the settlers weaker in numbers than themselves. When I arrived in Sungpan in 1910 I found there some 200 soldiers from Chengtu bent on a punitive expedition against these banditti. About a year previously a Chinese official had been murdered in the Amdo country, not many days' journey from Sungpan, and no redress had been obtained. Nine persons were held guilty for this crime, but in spite of demands on the part of the Chinese the clan would not give up these people. The affair ended in the Chinese killing as many members of this robber clan as the small army sent on the expedition could capture. It is from the Amdo region that Sungpan derives most of its wool, skins, and medicines, consequently the trade depends very largely upon peace obtaining there.

The Sifan (Western people) are unquestionably of Thibetan

origin. They are not nomads, but essentially a pastoral and agricultural people. In dress, speech, and facial characteristics they agree closely with the inhabitants of anterior Thibet. Their houses are similarly constructed, and Lamaism dominates their lives. As a people the Sifan are divided into several tribal clans : those around Sungpan style themselves Murookai ; those a little to the south-west of this town Lappä. Immediately around Sungpan the Chinese language is generally understood, but away from the town colloquial Thibetan only is spoken, each hamlet having an interpreter to conduct all affairs with the Chinese. These people are ruled by head-men who are held directly responsible for the proper maintenance of law and order. The Chinese policy is one of non-interference in so far as it is consistent with the status of China as the paramount power.

The Sifan men as seen in the streets of Sungpan and the immediate neighbourhood are swarthy in appearance and average 5 feet 6 inches in height or rather more ; in walking they have a clumsy gait and are generally awkward and sullen when approached. Their dress is a sort of " cover-all " made of grey or claret-coloured serge, confined around the waist by a girdle ; the right shoulder is generally uncovered. This garment is often edged with fur ; sometimes it is made entirely of sheep-skins, with the wool worn inside. Short trousers and high felt boots cover the legs and feet, though in the streets they frequently go barefooted. The head-gear is either a low, stone-coloured, soft felt hat, with turned-up brim bordered with black, or a high, cone-shaped, light grey felt hat edged with white sheep-skin. Occasionally those living near Chinese settlements affect a dirty turban. The hair is worn long and gathered up inside the hat. The Lamas have their heads close-cropped or shaven, and when seen in the streets are usually hatless. In ceremonial dress they wear a sort of cocked hat made of grey serge covered with a mass of fluffy yellowish woollen stuff. Muleteers and men generally, when travelling, go armed with swords, knives, and long guns, the latter fitted with a fuse and a fork to rest the barrel on when taking aim. All wear charm-boxes on their chests, and carry a flint-box and tinder suspended from their girdle ; somewhere

TOMB OF MAN MURDERED BY BANDITS ON HSUEH SHAN PASS

about their person a wooden, often silver-lined, eating bowl is also carried. The wealthy prize a leopard skin garment most highly.

The young girls are occasionally passing fair to look upon, but from hard work and exposure lose all charm of youth very early. The women are generally flat-faced, very dirty, and far from prepossessing. They have, however, considerable character and an important voice in household and all business matters. Toward foreigners they are timid, but amongst themselves their manners are playful, free and easy, and they laugh and sing at their work. Their outer dress consists of one shapeless piece of serge, which envelops them down to their ankles. Sometimes this is grey, more usually it is blue in colour, with a fancy bordering of dark red or yellow in front and around the bottom. High boots of untanned leather encase their feet and lower legs. Their hair is long and black, worn parted in the middle and collected into one large plait behind ; around the forehead it is worn in a series of tiny plaits ornamented with coral-beads, amber-coloured stones, and small shells. The large plait is usually wound around the head, together with a piece of cloth to form a kind of padded turban, the whole being decorated with shells and beads. Occasionally saucer-shaped felt hats are worn. In holiday attire, silver rings and gaudy red and yellow tassels are added to their coiffure. They are very fond of silver rings, bracelets, and large ear-rings ornamented with beads of turquoise and coral. In gala costume the dress is decidedly picturesque.

The men assist in tilling the soil, and in sowing and harvesting the crops, but the women do the bulk of the work around the homestead, the men being away herding the flocks or on journeys. Though they lead hard lives they seem a happy and contented people in spite of the fact that they are almost without exception afflicted with goitre. Their houses are built of wood and shale-rocks, being either one-storied, flat-roofed, with or without a raised part behind, or, as is more usual, two-storied and similarly roofed. They count their wealth in head of cattle, horses, and sheep. Wheat, barley, and peas are the staple crops. Meat, butter, and milk enter very largely into their diet. Buttered tea is generally drunk, but they are very partial to a kind of small beer which they brew from barley ; they are also fond of Chinese wine.

Monogamy is the rule, but polygamy is common, it being merely a question of wealth. Polyandry is not practised, but the morals are lax, as is the case everywhere else under Lamaism. Marriage is by consent on the part of the girl, presents of oxen and sheep being made on behalf of the bridegroom to the girl's parents ; children are appreciated, but the Sifan are not a prolific people. The second son generally enters a lamasery, as is customary throughout Thibet. Widows are permitted to remarry. The dead are disposed of by burial or by being thrown into the rivers.

Abundant signs of Lamaism are everywhere apparent. Prayer-flags flutter from the housetops, mountain-peaks, across streams, and surmount cairns of rocks. Mani-stones are heaped by the wayside ; praying-wheels, turned either by hand, by the wind, or by the currents of streams, occur on all sides. From the people at their work, either in low crooning tones or in loud chorus, the mystic hymn, " Om mani padmi hom," is continually ascending to heaven. The chant of the Sifan is decidedly musical, rising and falling in soft rhythmic cadence. I have often listened to them with much pleasure, though from a distance, since if one tried to approach closely they ran helter-skelter away. They are naturally very superstitious, being fond of charms, afraid of evil spirits, and reverence unusual natural phenomena. Though my associations with the Sifan were brief I always received the utmost courtesy at their hands, and found much that was pleasing and interesting among these happy, unsophisticated children of Nature.

CHAPTER XII

THE CHINO-THIBETAN BORDERLAND

"The Marches of the Mantzu"

IT is impossible to define, with any approach to accuracy, the political boundary between Szechuan and Thibet. Indeed, no actual frontier has ever been agreed upon, consequently it does not exist, except at one point, on the highway leading from Tachienlu, via Batang, to Lhassa. There, on the Ningching shan, three and a half days' journey west of Batang, stands a four-sided stone pillar, some 3 feet high, having been erected in A.D. 1728. The guide-book to Thibet says: "All to the east is under Peking; the territory to the west is governed by Lhassa." As to the regions north and south of this stone, nothing is said.

For all practical purposes the Min River, from Sungpan Ting in the north-west to Kuan Hsien, may be regarded as the frontier thereabouts. From Kuan Hsien southwards an imaginary line drawn through Kiung Chou, Yachou, Fulin to Ningyuan Fu, and thence to the Yangtsze River, may be accepted as completing the frontier line. This constitutes a well-defined ecclesiastical boundary between the peoples. Also it corresponds very closely with the western limits of the Red Basin, which constitutes an unmistakable physiographical frontier. It is true that at certain points, such as Lifan Ting, Monkong Ting, Tientsuan Chou, and Tachienlu, the Chinese have succeeded in establishing trading-centres and military depots. But in all these places the population is mixed and the centres themselves surrounded on two or more sides by non-Chinese people. West of the boundary here indicated the Chinese occupy a very limited aggregate area, being confined to the high roads and to a few valleys suitable for rice and maize cultivation. The largest of these areas is the region known as the Chiench'ang Valley, of which the city of Ningyuan Fu is the capital. This narrow strip extends down to the Upper Yangtsze River, being bounded on the east by the independent kingdom of Lolo, which occupies the higher slopes of the Taliang shan and has never been conquered by the Chinese. Immediately to the west of the valley the country is peopled by semi-independent tribes akin to the Thibetans. Indeed, the Min River, with such land to the immediate west suited to rice-culture, may well be regarded as the real boundary of western Szechuan from Sui Fu on the Yangtsze River to Sungpan Ting, in the extreme north-west corner of the province. An arc-line, commencing at Sungpan Ting and connecting with the boundary stone west of Batang, thence southwards, skirting the right bank of the Drechu (Upper Yangtsze), would form roughly the boundary of Thibet proper. Nominally the whole of this region is considered by the Chinese part of Szechuan province. In certain books and maps parts of this region are referred to as Eastern Thibet, and much confusion has arisen from this misnomer.

The country included within the boundaries here given constitutes the hinterland between Szechuan and Thibet, and failing a more lucid term it may be designated the " Chino-Thibetan borderland," a title which, if clumsy, has the merit of being both descriptive and accurate. Several trade routes traverse this borderland, but with one exception these have been little travelled by foreigners—the exception being the great highway between Chengtu Fu and Lhassa De, which crosses this region from Yachou, via Tachienlu and Batang to the boundary, and is closely controlled by Chinese. Apart from this highway and the country in its immediate vicinity as far west as Tachienlu, the whole borderland is very much a *terra incognita*. It is made up of a series of stupendous mountain ranges, separated by narrow valleys, well forested in the lower parts with all the higher peaks extending above the snow-line. These ranges are comparable only with the Himalayas, of which, indeed, they constitute a north-east extension. This rugged region is populated by many independent or quasi-independent tribes, more or less Thibetan in origin, with the exception of the Lolo.

THE HIGHWAY TO TACHIENLU AND BEYOND, HERE BLASTED
FROM SOLID ROCK

It is a region where altitude and climate, rather than longi-tude and latitude, define the frontiers.   In the north-west the highlands of Central Asia abut more closely on the Red Basin than they do in the south-west, and form uplands suitable as grazing-grounds for herds of yak, cattle, horses, and sheep. These areas are peopled by nomadic Thibetans, with whom agriculture is relatively unimportant.   The broken country, made up of mountain-crag and valley, which forms the greater part of this hinterland, is occupied by various tribes, with whom agriculture is the paramount industry, and wheat, barley, and buckwheat the staple food-stuffs.   The forests of this region contain much game, of which these people are skilled hunters.   Lastly, in the more fertile valleys, where rice and maize can be successfully grown, Chinese settlers are found, but, as mentioned earlier, away from trading-centres and the great highway between Chengtu and Tachienlu, they are not much in evidence

In the first chapter brief reference to the mountain chains and rivers of this region has been made, but perhaps a few of the more striking features may be given in detail here.   Unlike the mountains bordering the eastern limits of the Red Basin, which are mainly of hard Carboniferous and Ordovician lime-stones, those of the west are principally of mudshales and granitic rocks.   Here and there, for example Mount Omei and its sister mountains Wa and Wa-wu, hard limestones have been forced up through the older rocks and form bold peaks and stupendous precipices.   There is indeed plenty of limestone throughout the hinterland, but Pre-Cambrian rocks preponderate enor-mously.   These and the shales (probably Silurian) disinte-grate very readily in their exposed parts and erosion is rapid. In the deforested parts landslides are general.   The region is fairly rich in gold, silver, copper, lead, iron, and other minerals, but very little mining is carried on.   Coal is very rare, except in a few localities where limestone predominates, as near Mount Omei and the surrounding region.   Salt is known from one locality only (Pai-yen-ching in the Chiench'ang Valley).   Around Tachienlu hot springs of calcareous and chalybeate waters, more or less rich in sulphur, are common.   These springs are usually found in close proximity to torrents, very often occur-

ring in the actual bed of such streams. In many the waters
are actually boiling, and I have several times cooked eggs in
them. These hot springs are much resorted to by the people
of the surrounding regions for bathing purposes, the waters
being esteemed as a cure for rheumatism, skin affections, and
other complaints.

Three large rivers, Tung, Yalung, and Dre, flow through
this borderland, mainly from north to south, as necessitated by
the direction of the mountain axes. These rivers have tributaries in abundance, and the majority of them, draining from
eternal snows, carry down enormous quantities of water and
detritus. None of these rivers is navigable save for rafts,
specially constructed boats, or skin coracles, over very short
and interrupted stretches. Bridges and ferries are few, nevertheless the highways and by-ways of this region skirt the banks
of these rivers and their main tributaries.

The valleys of all these streams, and for the purpose of
what follows the Min above Kuan Hsien may be included, are
deeply eroded, the waters flowing between steep slopes or precipices. These valleys are all very similar, being narrow, shut
in by lofty treeless mountains, and all enjoy a much hotter,
drier climate than their altitude warrants. Long stretches
are very barren and desert-like, more especially when the outcropping rocks are solely granitic. Owing to this dry, hot
climate, interesting anomalies obtain in these valleys. At
Hokou, on the Yalung, maize can be cultivated up to nearly
9500 feet altitude, whereas at Tachienlu, in the same latitude
and 1000 feet less altitude, it is impossible to bring this cereal
to maturity. Green parrots (*Palæornis derbyana salvadori*)
occur as summer migrants in the valleys of the Yalung and
Drechu up to 10,000 feet altitude. Rock pigeons occur in
multitudes throughout all these valleys above 4000 feet altitude.
Monkeys also are common. The flora generally is specially
adapted to withstand drought, and is more closely allied to
that of the Yunnan plateaux than to the contiguous country.
Doubtless at one time the mountain-slopes flanking these
valleys were wooded, though it is improbable that the lower
slopes were ever heavily forested; but such timber as grew
there has long since disappeared, and to-day these slopes are

clothed only with coarse grass and scrub. Landslides are a feature of these regions, especially during the melting of the snows or after heavy rains in the surrounding high mountains. At such times travelling hereabouts is highly dangerous, as nearly every traveller can testify from ocular proof. I have witnessed several disastrous landslides, involving loss of life and much destruction of property. In 1910, when descending the Min Valley, I unfortunately got involved in a minor one, and sustained a compound fracture of the right leg just above the ankle. In many places rockslides are constantly occurring, and warning notices to travellers not to tarry are frequently displayed throughout the Upper Min Valley and elsewhere.

Small villages and farmsteads are scattered through these valleys where, goaded by stern necessity, the inhabitants maintain a grim struggle to win a sustenance from the inhospitable soil. Where rice and maize can be cultivated Chinese settlers are found, but above the altitudes admitting of this the tribes are in full possession and cultivate crops of wheat, barley, buckwheat, peas, and linseed—the latter for its oil, which is used as an illuminant. Exceptionally good Chilli peppers (*Capsicum*) are grown in these valleys, and certain regions, for example Mao Chou, in the Min Valley, are renowned for this produce. Around habitations a few trees, chiefly Poplar, Alder, and Willow, are always present, affording a welcome shade. *Cupressus torulosa*, a handsome timber tree, often 80 to 100 feet tall, is very much at home in these valleys and probably at one time covered quite considerable areas hereabouts. This tree is well worth the attention of those engaged in reafforestation work in dry, warm-temperate regions. Other trees partial to these same conditions are *Sophora japonica, Diospyros Lotus, Pistacia chinensis, Erythrina indica, Kœlreuteria apiculata, Ailanthus Vilmoriniana*, Celtis spp., and the Soap trees (*Sapindus mukorossi*, Gleditsia spp.). Many fruit trees occur, including the Pear, Apple, Peach, Apricot, and Walnut; the latter (*Juglans regia*) is the commonest tree up to 8000 feet. The natives hack the lower trunk to make the tree fruitful, so they claim, showing that the old adage—" beating the Walnut tree "—is known outside of Europe. Mulberry trees, the *Cudrania tricuspidata*, and tall-growing Bamboos are common up to 4500 feet altitude.

Many of the shrubs found growing in these valleys are spinescent and nearly all are adapted to withstand drought. In the majority the leaves are very small or covered with a dense felt of hairs.   These shrubs are usually scrubby in appearance yet many produce ornamental flowers or fruit.   The " Southernwood " (*Artemisia* spp.), with silver-grey, elegantly dissected foliage and yellow flowers, are perhaps the commonest shrubs met with hereabouts.   Barberries are another special feature, and when laden with masses of red fruit and autumn-tinted foliage present a most attractive picture.   This same remark applies to various species of Cotoneaster, all having ornamental fruit.   Many kinds of Rose occur, but often the species are local.   Common to all these valleys, though most abundant in that of the Yalung, is *Rosa Soulieana*, with fragrant flowers, opening sulphur-yellow and changing to white.   So also is Miss Willmott's charming rose (*R. Willmottiæ*), with its abundant straw-yellow prickles, neat glaucescent leaves, rosy-pink flowers, and orange-red fruit.   The beautiful *R. Hugonis* is confined to a narrow stretch of the Min Valley between 3000 to 5000 feet.   This is the only rose with yellow flowers I have met with in China.   The fruit is black and falls very early.   *R. multibracteata*, an odd-looking species having pretty pink flowers, is very common in the upper reaches of the Min Valley and less so in that of the Tung. Forms of the Musk Rose (*R. moschata*) and of *R. sericea* occur but are local.   With the exception of the " Southernwood," all these shrubs confine themselves closely to watercourses.   In more arid places *Caryopteris incana* and other species, with intense blue flowers opening in late July, are very abundant, so also are different species of Indigofera, with pink to red-purple flowers.   Several species of Buddleia and two varieties of the lovely *Clematis glauca*, with glaucous foliage and top-shaped, yellow, passing to bronze-coloured flowers, ought not to be overlooked.   The shingly and sandy foreshores are covered with Willow, Sallowthorn, and False Tamarisk (*Myricaria germanica*).   In the Tung Valley, between

A PRICKLY PEAR (OPUNTIA DILLENII) NATURALIZED

4000 and 5000 feet altitude, a " Prickly Pear " (*Opuntia Dillenii*) has become naturalized. This American colonist has made itself very much at home, covering many miles of barren rocky slopes. It grows 6 to 10 feet tall, and when covered with its yellow or pale orange flowers is very ornamental. The edible nature of the fruit is well known to the natives but is little esteemed. An extract obtained by boiling the fleshy stems is locally employed as a supposed cure for hæmorrhoids.

Amongst the coarse grass and scrub, the dominant features of these regions, a variety of showy herbs occur, nearly all having bulbous or thickened rootstocks in some form or other. To garden lovers everywhere these valleys are of special interest, inasmuch as they are the home of many beautiful Lilies. Each of these valleys has species or varieties peculiarly its own, which range up to about 8000 feet altitude, yet whilst very local these Lilies are numerically extraordinarily abundant. In late June and July it is possible to walk for days through a veritable wild garden dominated by these beautiful flowers. In the Min Valley the charming *Lilium regale* luxuriates in rocky crevices, sun-baked throughout the greater part of the year. It grows 3 to 5 feet tall, and has slender leaves crowded on stems bearing several large funnel-shaped flowers, red-purple without, ivory-white suffused with canary-yellow within, often with the red-purple reflected through, and is deliciously fragrant. In the Tung Valley, Mrs. Sargent's Lily (*L. Sargentiæ*), a taller growing species than the foregoing, with broader leaves, having bulbils in the axils, equally handsome flowers of similar shape, but varying from green to red-purple without and from pure white to yellow within, is very abundant in rocky places among grass and scrub. The flowers of this species are collected, boiled, and dried in the sun, then minced, fried with salt and oil, and eaten in the same way as preserved cabbage. The bulbs of the Tiger Lily (*L. tigrinum*) and its elegant ally, *L. Thayeræ*, which are white, are cooked and eaten. Several other Lilies abound in these valleys, including the lovely *L. Bakerianum* and other species not yet named.

A herb very common in the Tung Valley is *Thalictrum*

dipterocarpum. This Meadow-rue grows 6 to 8 feet tall, has elegant, much-divided foliage, and multitudinous, large, lavender-purple flowers—by common consent the handsomest member of its family. In the Min and Tung Valleys, but very local, *Incarvillea Wilsonii*, which grows nearly 6 feet tall and has handsome flowers very like those of *I. Delavayi*, occurs. This plant is monocarpic and has not yet flowered in cultivation, although I introduced it into the Veitchian nurseries as long ago as 1903. *Salvia Przewalskii*, with large purple flowers, is another striking herb common in the valleys above 8000 feet altitude. This list of ornamental herbs could easily be extended if any useful service would be served thereby. On bare rocks various species of Selaginella abound; the Mullein (*Verbascum Thapsus*), Deadly Nightshade (*Hyoscyamus niger*), and Thornapple (*Datura Stramonium*) are common weeds by the wayside. The poisonous properties of the two last named are well known to the natives. From this brief sketch it will be seen that these narrow, dry, almost desert-like valleys, with their abnormally warm climate, possess a flora which, if limited in number of species, contains many plants of more than passing interest and horticultural value.

As mentioned earlier (p. 149), this hinterland is peopled by various independent and semi-independent tribes about which little is known. The whole region is analagous with that separating India and Thibet, and this statement of fact will perhaps convey a more intelligible idea than the most voluminous details. These tribes are divisible into four distinct groups, in accordance with their official status and form of government.

1. States independent, non-tributary, hostile to both Chinese and Lama authority, as the Lolo kingdom. I have no intimate acquaintance with the Lolo—a people once spread over much of Yunnan, but now relegated to the region of the Taliang shan, where they have never been conquered by the Chinese. This race possesses a written language peculiar to itself and is probably indigenous.

2. States really independent and even hostile toward China, directly controlled by the Dalai Lama and Council, whose policy is supposed to be modified by High Commissioners appointed

by China, as, for example, Chantui, Derge, and Sanai. The territory occupied by these tribes is west of the Yalung River and contiguous to that of Thibet proper ; the people are indistinguishable from those inhabiting anterior Thibet generally. These more western regions have been styled the " Thibetan Marches." Some four years ago, an acting viceroy of Szechuan, one Chao Êrh-fêng, was appointed Warden of these Marches. With an army of Chinese soldiers he indulged in a most aggressive policy and speedily subjected the whole region to Chinese control. He broke the Lama power, destroyed the principal lamaseries, and beheaded the abbots and other dignitaries. His task was rendered fairly easy owing to affairs in Lhassa, consequent upon the British expedition to that city, and the flight of the Dalai Lama, the whole making impossible any concerted action by Lhassa De in support of their adherents in the Marches. (In 1911 Chao Êrh-fêng was appointed Viceroy of Szechuan and was subsequently murdered in Chengtu city by Chinese revolutionists.)

3. States tributary-controlled, governed by hereditary native princes and subject to the Viceroy of Szechuan in temporal affairs, but more or less strongly influenced by the Dalai Lama, owing to Lamaism being the accepted religion. Of these the kingdom of Chiala, the Horba states, and the Chiarung tribes are the chief. They occupy most of the territory between the Min and Yalung Rivers north of a line connecting Yachou with Tachienlu and Hokou. The Chiala kingdom I shall deal with separately when describing Tachienlu, the capital city. The Chiarung are dealt with in the next chapter.

4. A number of very small states, governed by quasi-independent chiefs, indirectly controlled by Chinese officials appointed for that purpose and by the surrounding tributary kingdoms. They are, in fact, tiny buffer areas very useful to the Chinese in maintaining the balance of power among the larger, more independent kingdoms. Many of these principalities are made up of people who may reasonably be looked upon as remains of the aboriginal population of parts of Szechuan and this hinterland. These petty states are scattered through the more easterly parts of this hinterland

from Mao Chou in the north through the Chiench'ang Valley to borders of Yunnan province. The power exercised by the chiefs varies according to their proximity to thickly populated Chinese districts or otherwise. In the former it is almost nominal, whereas in the latter case it is very considerable.

In addition to the above are certain feudal states whose overlord owes his office directly to Chinese influence, and who is bound, if called upon, to render military service to China. These feudal chieftainships are hereditary and were originally bestowed as rewards for assistance rendered to the Chinese in breaking up the Chiarung confederacy during the reign of the Emperor Kienlung. Many of these, for example the Tsa-ka-lao chief, have very considerable power and influence in the temporal affairs of the surrounding tributary-controlled kingdoms. The people are mainly of the same stock as the Chiarung tribes. All the chiefs of these feudatory states and tributary kingdoms are closely related by inter-marriage.

The Chinese designate the inhabitants of this borderland " Mantzu," a contemptuous term signifying " Barbarian " and of no ethnological value whatever. But the policy they have pursued in dealing with these people has been shrewdly wise if unscrupulous. With arms and money the Chinese have displayed their power and obtained what practically amounts to a suzerainty over the whole borderland. A former emperor said : " Wardens of the Marches should seek to checkmate the native tribes by becoming intimately acquainted with them and their customs and thus able to prevent any united action. In this way the tribes will remain weak and easy to manage. They should be encouraged to appeal to Chinese authorities for advice and protection in their disputes with one another. These authorities will, of course, be in no hurry to settle their cases. If the tribes are taught to fear the Chinese, and the officials act with energy, all trouble will be avoided." This crafty advice has long been acted upon by the Chinese, with much success from their own view-point.

From this brief and very incomplete general account it may be gathered that this hinterland is a fascinating region, presenting ethnological and other problems of great interest, the solution of which is worthy of the attention of Western scientists. It is hoped that a properly equipped expedition will at no very distant date be organized and dispatched to survey and investigate fully this little-known Chino-Thibetan borderland.

CHAPTER XIII

THE CHIARUNG TRIBES

Their History, Manners, and Customs

WITHIN the limits of the Chino-Thibetan borderland, as defined in Chapter XII, from Sungpan Ting southwards to Yachou Fu, and west to the valley of the Upper Tung or Tachin (Great Gold) River, the territory is divided amongst numerous cognate tribes collectively spoken of by Chinese as " Chiarung." These people are essentially agriculturists, making their homes in the upland valleys. They are all, though tributary to China, ruled by their own hereditary chiefs ; each tribe occupies a properly defined area, with its own capital town, the political centre of the entire region being Monkong Ting. These tribes are non-Chinese and are not indigenous to this region. They are also distinct from the people found in anterior Thibet. They speak a difficult and at first sound unpronounceable jargon, which, if it be the mother of Thibetan dialects, is widely different from that spoken in Thibet to-day. But Thibetan letters have, without difficulty, been applied to it, and scholars, priests, officials, and merchants both read and speak the Lhassa-Thibetan language with greater or less fluency.

The origin of these people is obscure, yet there is good reason to believe they come originally from the region around the head-waters of the Tsang-po (Upper Brahmaputra River), and probably have common origin with the people of Nepal and Bhutan. Personally, I am of the opinion that they came over with Genghis Khan, or his son Ok-Ko-Dai, at the commencement of the thirteenth century, and assisted in the conquest of western Szechuan. As a reward for military services rendered

THE TACHIENLU RIVER

they were given the territory they occupy to-day. During the course of time they waxed powerful, menacing the territory to the east of the Min River, and even taking possession of certain parts. In Ming times the Chinese made war with them on many occasions. They were a source of trouble to the Manchu dynasty until the famous Emperor Kienlung determined upon crushing their power. After a very fierce struggle this was accomplished by a Chinese general named A-kuei. First he subjugated the region of the Hsaochin Ho (Little Gold River), then, after much difficulty, he captured Lo-wu-wei (modern Hsuching), the capital of the Tachin Ho (Great Gold River), took the king prisoner, and made a map of the entire region. The king, named Solomuh, was sent to Peking, where, after a grand court ceremony, he was sliced to pieces. The conquest was completed early in A.D. 1775. Military colonies were then established by the Chinese in strategic places, the more fertile regions were confiscated, and Chinese settlers induced to take possession. In crushing this confederacy the Chinese were assisted by the tribes, being to some extent divided amongst themselves. Some of them fought on the Chinese side, and as a reward certain areas situated at strategic points were fiefed out and established as feudal states for the benefit of these allies, an overlord with hereditary control being appointed to each. The Chinese handled this campaign with consummate skill, and the administrative system established has remained unchanged down to the present day. The power of the tribes was completely broken ; and the feudal states and the military colonies have safeguarded the Chinese from any concerted action on the part of these people ever since. It will, however, be readily understood that the tribes farthest removed from regions fully occupied by Chinese enjoy to-day greater independence than those in close contiguity.

Originally these " Chiarung " had one common language, but time, isolation, and the dividing up into clans has produced many very dissimilar dialects. These people are now split up into eighteen tribes, occupying very unequal areas of territory, and though all are interrelated by marriage they are by no means at peace with one another. Feuds are constant, and fighting among themselves is very much the rule. Since

this keeps them weak in power the Chinese policy is to
intervene as seldom as possible. On the map are indicated as accurately as our knowledge admits the positions
occupied by some of these tribes and feudal states. It is
almost impossible to render into English the guttural sounds
denoting the names of many of these tribes. But, fortunately,
the more important, namely, Mupin, Wassu, Somo, Damba,
Bati-Bawang, Wokje, are the least difficult to pronounce.
The whole territory occupied by these people is about 250
miles from north to south, and 200 miles east to west at broadest
point. The population is about half a million.

Two main roads, one from Kuan Hsien, the other from
Lifan Ting, cross this region and unite near Monkong Ting.
In addition, a network of cross-country by-ways connects
the various villages and states.

The Chiarung are essentially agriculturists, cultivating
with much skill crops of wheat, barley, peas, buckwheat,
maize, Irish potato, and miscellaneous vegetables. Sheep,
cattle, ponies, and goats are kept by the more wealthy, often
in quantity. The horses are sold to Chinese traders, but the
wool is woven into cloth for their own use. Milk, butter, and
meat enter largely into their diet. They are also skilled gun-
and swordsmiths, more especially the Somo people, who manu-
facture most of these weapons in use among the tribes themselves and the people of eastern Thibet generally. Many are
also highly skilled masons, builders, and well-sinkers, and as
such have a reputation even amongst the Chinese. During
August many visit the upper reaches of the Min River every
year to take part in harvesting the crops ; indeed, for this
purpose they supply most of the labour in that region. Often
they are in request in Chengtu and other cities for sinking wells
and such-like work.

The Chiarung live in settlements of from several to a
hundred families or even more, always in positions admirably
suited for defence. These settlements usually crown some
bluff or eminence ; very often they are perched like an eagle's
aerie high up on the steep mountain-side. The architecture
which obtains throughout is characteristic and peculiar. Each
settlement is dominated by one or more tall, chimney-like

THE WOKJE VILLAGE OF TA-WEI

towers, either square, hexagonal, or octagonal in shape, 60 to 80 feet high, and resembling from a distance the stack of some large factory in Western lands. The exact significance of these towers it is difficult to fathom, but it is evident that they can serve as storehouses, watch-towers, and harbours of refuge in times of stress and war. They have also some obscure connexion with religious matters, possibly in this they have some remote affinity with the pagodas of China and Burmah. The houses are more or less square, flat roofed, solidly built of shale-rock and mud. Those belonging to the chiefs and men of property are three or four stories high. The walls are thick, pierced with loopholes and several narrow latticed windows. At all four corners of the roof turrets 3 to 4 feet high are built, sometimes there are more in different patterns. From these prayer-flags are displayed, often with the green branches of Juniper. On the roof also is fixed an incinerator for the sacrificial burning of fragrant juniper branches as incense. Part of the roof is frequently occupied by a hurdle-like framework called " Kai-kos," 10 to 15 feet high, which is employed for drying grain upon. The rest of the roof is used for religious exercise, eating, sleeping, and recreation ; in harvest-time it serves as a threshing-floor. The ground story is made up of a courtyard surrounded by sheep and cattle-pens, the kitchen, and usually a guest-room.

The turrets, upper rim of the walls, edges of the window-spaces, base and base angles of the walls, are washed white, commonly white lines stretch diagonally up the walls, and the swastika cross, with other devices and symbols, are displayed in white on these walls. Crowning the edges of the roof, or arranged on separate structures, symbols denoting a globe, upturned crescent, and the swastika are commonly displayed. The lamaseries are similarly constructed, only larger, and usually with more stories. The houses of the peasants also are on the same plan, but of one or two stories only. All these structures are closely packed together with one to several towers reared above the whole assemblage. The different emblems and symbols of Nature worship may occur in the structure of Thibetan houses and lamaseries, but the tall tower is peculiar to the Chiarung.

Another interesting feature of these regions is the bridges. All these structures are of designs differing from those found throughout China proper, but agreeing closely with those in use throughout Sikhim, Bhutan, and Nepal, thus furnishing additional evidence of the affinity of these peoples. All the smaller streams and torrents are bridged by logs arranged on a semi-cantilever principle, and call for no special remark. But the larger streams are crossed by suspension bridges constructed of split and plaited bamboo cables. These bridges are very similar to the cane bridges of Sikhim and Bhutan. They are found throughout the territory occupied by these tribes and the narrow strip of territory wedged in between the Min Valley and the western limits of the Red Basin. This latter strip was formerly occupied by these tribes, and is to-day largely peopled by their descendants or half-caste Chinese. As mentioned in Chapter X, pp. 117, 130, iron suspension bridges occur in one or two places in the north-west corner of Szechuan. This style of bridge is common from the valley of the Ya River, and the Tung at Luting chiao, southward to the frontier of Burmah, and is probably of Shan origin. Similar bridges of iron rods and chains are met with in Bhutan, where they are considered to be of Chinese origin (White, *Sikhim and Bhutan*, p. 191). Throughout the Chino-Thibetan Borderland iron and bamboo are equally common, yet it is a singular fact that their use in bridge-building is restricted to definite areas.

Cable or rope bridges are abundant throughout the entire region, and extend much farther west and south than the Chia-rung territory. These simple but extremely useful structures consist of a bamboo hawser stretched across the stream usually from a higher to a lower point; if the stream is moderately narrow the question of incline is of less importance. The hawser may be anything from 8 inches to 1 foot thick, and being heavy sags considerably in the middle, unless the stream is very narrow, as around Tachienlu, where a rather different method of crossing than that about to be explained is in vogue. To cross one of these cable bridges a person is supplied with a length of strong hempen rope hanging free from a saddle-shaped runner of oak or some other tough wood. The runner clips the cable, and the hempen rope is fastened under and around the legs and waist to form a "cradle." When all is properly secured the person throws one arm over the top of the runner, gives a slight spring, and glides down the inclined cable at increasing speed. The impetus obtained in the downward rush carries the passenger over the central dip and more or less up the lesser incline on the opposite side. If the momentum is insufficient to land the person, the remaining distance has to be traversed by taking hold of the hawser and hauling hand over hand. Crossing these bridges is fearsome work until one is accustomed to it. It is speedily accomplished, and there is practically no danger so long as one keeps a cool head and the ropes do not break. It is a common sight to see men with loads and women with children on their backs cross these bridges. But heavy loads are usually fixed to the runners and hauled across by a rope attached to them.

None of the rivers traversing Chiarung territory is navigable in the ordinary sense of the term, but skin coracles, broadly oval in shape, descend certain stretches of the Upper Tung River. These frail boats serve also to ferry over goods and passengers at certain necessary places. They are made of cattle hide stretched over ribs of tough, light wood. The whole coracle is very easily carried by one man, and closely resembles pictures of the boats used by ancient Britons prior to the Roman invasion. They are steered by a man seated in the stern operating a paddle, and accommodate about two passengers. A passage down or across stream in one of these coracles consists very largely in describing, more or less rapidly, a series of wide circles and half circles. As a novelty, productive of excitement, not unmixed with danger, these coracles and cable bridges can with confidence be recommended to "World's Fair" promoters and showmen generally. The skin coracle is in general use at ferries throughout Eastern Thibet and the Marches, and is not strictly a Chiarung specialty.

In height the tribesmen average about 5 feet 7 inches or rather more; the face is usually oval, with rather pointed chin, straight nose, sometimes inclining towards aquiline. They dress ordinarily in undyed serge cloth of local make, worn in the same manner as that of the Sifan. The legs are swathed in felt putties; the head-gear is either a turban

or black pudding-basin-shaped felt hat.   Those living near
Chinese settlements and the highways have their head in part
shaven, and wear their hair in a queue Chinese fashion.   On
holiday occasions their garments are brightened with red
bordering, and high felt boots are worn.   The women are
short in stature (about 5 feet), sturdy and buxom, somewhat
gipsylike, with dark olive complexions, and when young are
often good looking.   Their ordinary outer dress is a garment
of grey native serge of no definite shape, reaching to just
below the knee and bound around the waist with a scarf.
The legs and feet are bare or encased in top-boots.   Commonly
they go bare-headed with their long black hair parted down
the middle and hanging down the back in one large plait.
They are fond of large bangles, ear-rings, etc., made of silver
inlaid with turquoise and coral.   On festive occasions
garments edged with red and very often made of blue cloth
are worn.   The more wealthy dames decorate themselves very
lavishly with silver ornaments, and wear covering their heads
a piece of cloth held down by means of their large plait of hair,
which is wound around and decorated with silver and beads
of coral and turquoise ; the lower part of the piece of cloth
hangs free over the back of the neck and shoulders.   These
dames are women of character, and have a ruling voice in
household and family matters generally ; also, from what I
saw of them, they appear to conduct most of the business.
These women lead a strenuous life ; they cultivate the fields,
tend the flocks, take the farm produce to market, hew wood,
and carry water.   The domestic duties of cooking, making
and mending clothes and general household work devolve
upon the men.   Yet the women are not unkindly treated,
and are far from being down-trodden.   Being of cheerful
disposition, they seem well suited to the free outdoor life they
lead, and laugh and sing as they ply their task.   Among
themselves these people are frank and easy in manner, and the
women enjoy a freedom of position unknown amongst the
Chinese.   A party of dames and men were fellow-travellers
with me once for a couple of days.   When the time came to
separate they made merry over cups of wine ; the women
officiated, and cordially invited me to join them.   With their

IDOLS IN A BÖNPA TEMPLE

laughter and song they made cheery companions, and I was sorry to part from them.

The families are small, but the children are usually strong and healthy. Girls marry between the ages of seventeen and twenty, polygamy is common, but polyandry is unknown except, perhaps, in the upland regions bordering Thibet proper. Temporary marriages, so general in Thibet, are also unknown amongst the Chiarung. Nevertheless, the standard of morals in vogue among these people is a very low one. In certain states hetairism precedes maternity. In Badi-Bawang the unmarried girls and childless women wear only two sporran-like fringes of woollen threads or pieces of fur, suspended from a girdle passed around the body above the hips. The legs are exposed, but the upper parts of the body are usually covered by a coarse serge garment. Only after their first child is born may they wear skirts, since the gods have then purified them. A pregnant damsel selects from among her lovers a husband, who thus becomes the accepted father of her child, her word in this matter being final. Maternity alone ratifies marriage, and indeed saves women from promiscuity. The defloration of virgins is the prerogative of chiefs and head-men, but is not always exacted. In many ways these people are apparently shameless, according to Chinese and Occidental ideas alike. It is no uncommon sight to see women of all ages, quite nude, bathing in streams by the wayside. This same custom is also common at Tachienlu, where the hot springs are favourite bathing-places for both sexes. But after maternity the women are said to remain constant ; divorce or legal separation after ratified marriage are not practised.

The explanation of the above and other curious customs of these interesting people is found in their religious beliefs. Although orthodox Lamaism is more or less paramount the mysterious Bönpa religion, with its marked tendency toward phallic worship, lurks throughout the lonely valleys of the Chiarung tribes. In Badi-Bawang it is the recognized state religion. It should also be remembered that these regions constituted the famous matriarchal kingdoms of Chinese historians. Indeed, even to-day, certain states have queens holding nominal or actual authority, and in these in some capacity a woman must always rule. Occasionally the difficulty is overcome by styling the ruling head a " Queen " quite irrespective of sex !

Lamaism appears in three forms, the Yellow, Red, and Black, the latter representing the Bönpa cult. The religious centre is Tsong-hua on the Tachin River, about 60 miles west of Monking Ting. But lamaseries are scattered over the land, occurring separately by themselves or in association with the residences of the hereditary chiefs. The Yellow or orthodox sect is first in importance and numbers, and is controlled directly from Lhassa. The ritual differs in no way from that practised throughout the hierarchy of Lamaism. The same remark applies to the unorthodox Red sect, which is of much less importance, and whose priests are allowed to marry.

The Black or Bönpa sect has a ritual bearing an outward resemblance to orthodox Lamaism, but apart from this there is little else in common. In many things the Bönpa are the avowed enemies of the orthodox. They turn their praying wheels from left to right instead of from right to left ; they pass sacred objects on the right instead of on the left ; also they refuse to repeat the mystic Mantra, " Om mani padmi hom," replacing it with one peculiarly their own. As to the origin of this Bönpa it is difficult to say. My friend, Mr. J. Hutson Edgar of the China Inland Mission, who has travelled among and studied these Chiarung tribes more closely than any one else living, inclines to regard it as the remains of the old Nature worship of Thibet, which probably underlies all the religious systems of Eastern Asia.

In the state of Wassu are several temples belonging to this Bönpa sect. Through the courtesy of the chieftain I was allowed to inspect some of these temples, and succeeded in obtaining fair photographs of the idols. These latter, made of stone, wood, straw, and plaster, represent giants and demons with their female energies ; the walls are decorated with paintings depicting erotomania. Hideous and disgustingly obscene are the contents of these temples, where phallic worship holds unblushing sway. The Wassu chief informed me that the Mantra used by these Bönpa priests is " Hom ma-te ma-tsi ma-yöor tsa-lien doo." He kindly gave me a copy of this hymn, but I have not yet succeeded in getting it translated into intelligible English. The principal symbol in use is the Fylfot or swastika, which they call " Yungdrung." A mystical bird, " Chyong " or " Garuda," is also regarded with great favour as an emblem of fruitfulness. In the Bönpa temples at Tung-ling shan, near the residence of the Wassu chief, I also recognized the image of Kwanyin (Goddess of Mercy), the God of Wealth, and many demons similar in appearance to those found in ordinary Buddhist temples throughout China proper. It would thus appear from the catholic nature of the contents of their temples that these people accept a measure of Buddhism, and Lamaism both orthodox and unorthodox, and the Bönpa in its entirety. An atmosphere of secrecy and mystery enshrouds the Bönpa temples, which are frequently built in places difficult of access. The cult has been subjected to much persecution at the hands of Lamaists, yet, notwithstanding, it retains a firmer hold on the people of most of the Chiarung states than any other form of religion. In their hearts children of nature, their daily life one constant struggle against an inhospitable soil and climate to win a crop necessary for their sustenance, these people very naturally incline most toward the gods of Increase and Fecundity.

CHAPTER XIV

ACROSS THE CHINO-THIBETAN BORDERLAND

KUAN HSIEN TO ROMI CHANGO; THE FLORA OF THE PAN-LAN SHAN

URING the summer of 1908, when in Chengtu, I determined upon a journey to Tachienlu. Previously, in 1903 and again in 1904, I had visited this town by three different routes. This time I decided upon following the road leading from Kuan Hsien via Monkong Ting and Romi Chango. The only published account of this route that I have knowledge of is in a Report by Mr. (now Sir) Alexander Hosie,[1] erstwhile H.B.M.'s Consul-General at Chengtu, who returned from Tachienlu over this road in October 1904. What is written in this Report about the forests of that region created a desire within me which nothing short of actual experience could satisfy. Again, this route promised further acquaintance with the tribesfolk inhabiting the hinterland. Sir Alexander's description of the road portrayed a difficult journey, but I felt sure that by taking time and but lightly burdening my men I could get through all right. This confidence was fully justified, as events proved, and what I saw of the forests and mountain scenery, together with the quantity and variety of plants discovered and collected, abundantly repaid me for the hardships experienced. The journey is estimated at 1326 li, approximately 330 English miles, but, whilst mere mileage is of little moment in mountainous countries, I should consider 250 miles a more accurate figure.

With Tachienlu as my goal I left the city of Chengtu on the morning of 15th June, and at noon the next day reached

[1] *Journey to the Eastern Frontier of Thibet*, presented to both Houses of Parliament, August 1905.

A BAMBOO SUSPENSION BRIDGE 70 YDS. LONG

the city of Kuan Hsien. An afternoon sufficed to complete my arrangements. The caravan consisted of eighteen carrying coolies and one head coolie, two chairs, two handy men, an escort of two soldiers, my Boy, and self, making a party of thirty all told. The journey occupied twenty-three days from Kuan Hsien.

What follows is compiled from my diary :—

The famous bamboo bridge, known as the An-lan chiao, over which the road to Monkong Ting passes, was having its annual overhauling; in consequence, on leaving Kuan Hsien we had to journey down stream some 5 li to a point where it was possible to cross the various arms of the Min River by improvised bridges and ferry. In so doing we had an opportunity of realizing, somewhat hazily be it confessed, what this area must have been like before Li-ping's wonderful irrigation works came into existence. Without counting the streams flowing Chengtu-wards we crossed five distinct arms of the Min River proper scattered over an area a mile wide, covered with sand, shingle, and coarse grass (*Miscanthus sinensis*). The detour involved 15 li, and it was not until 9 o'clock that we were opposite the An-lan chiao. This most remarkable structure is about 250 yards long, 9 feet wide, built entirely of bamboo cables resting on seven supports fixed equidistant in the bed of the stream, the central one only being of stone. The floor of the bridge rests across ten bamboo cables, each 21 inches in circumference, made of bamboo culms, split and twisted together. Five similar cables on each side form the " rails." The cables are all fastened to huge capstans, embedded in masonry, which are revolved by means of spars and keep the cables taut. The floor of the bridge is of planking held down by a bamboo rope on either side. Lateral strands of bamboo keep the various cables in place, and wooden pegs driven through poles of hard wood assist in keeping the floor of the bridge in position. Not a single nail or piece of iron is used in the whole structure. Every year the cables supporting the floor of the bridge are replaced by new ones, they themselves replacing the " rails." This bridge is very picturesque in appearance, and a most ingenious engineering feat.

From the An-lan chiao the road ascends the right bank

of the Min River, and is broad, in good repair, but with many awkward gradients. We found lodgings for the night at Hsuan-kou, alt. 2640 feet, a market village of some 300 houses, situated on a tributary immediately above its union with the main stream, which describes a very sharp turn on leaving a narrow gorge. The Min River from the An-lan chiao to this point is full of minor rapids, and the current is very swift. Near Hsuan-kou timber is made into rafts and floated down to Kuan Hsien, thence to Chengtu and elsewhere.

During the day's march we passed some good-sized trees of Black Birch, Nanmu (*Machilus* spp.), Hog-plum (*Spondias axillaris*), and small trees of *Cryptomeria japonica*, the latter obviously planted. A large trumpet-flowered Lily was abundant in rocky places by the wayside. Rice occurred sporadically, but the principal crop was maize. Around the inn Tea-bushes are abundantly planted.

On leaving Hsuan-kou we crossed the tributary by a small bamboo suspension bridge, and ascended the left bank by an easy road for 30 li to Shui-mo-kou. Throughout this stretch Cryptomeria is common. All the trees are small and obviously planted, yet I cannot rid myself of the idea that it must be indigenous somewhere in this vicinity. It occurs scattered over a large area, always near habitations, yet it is scarcely feasible to suppose that this tree has been brought from Japan for the purpose of planting it hereabouts.

Shui-mo-kou is an ordinary Chinese market village of some 350 houses lining either side of the main street. It is interesting, however, as being the last purely Chinese village in this direction, also the last place wherein supplies can be purchased or silver exchanged until Monkong Ting is reached. I hired an extra man, and all my followers laid in a stock of rice and food-stuffs generally. At Kuan Hsien, appreciating fully the difficult road before us, I had reduced all loads to two-thirds the normal weight. In spite of this the carriers were heavily laden with extra supplies, and could hardly stagger along on leaving Shui-mo-kou.

A short distance beyond the above village there is a steep ascent, but after a few li the road becomes easy and winds around the mountain-side. Scrub Oak and unhappy-looking

THE VILLAGE OF HSUAN-KOU

trees of Cunninghamia are abundant, but the flora generally is poor. Wild Strawberries cover the more grassy slopes, and were laden with white and red luscious fruit. We passed a few houses, and finally reached the top of the ridge, alt. 5600 feet, which is known as the Yao-tsze shan. Crossing over we entered the territory under the jurisdiction of the Wassu chieftain, who resides at Tung-ling shan, near Wênch'uan Hsien in the Min Valley.

Descending by a path, which at first easy soon becomes very precipitous and difficult owing to the abundance of loose rocks, we reached Hei-shih ch'ang, our destination for the day, at 6 p.m. In this descent, near the head of the pass, the "Yang-tao" (*Actinidia chinensis*) is abundant, and was laden with a wealth of large, white, fragrant flowers. By the wayside, *Rosa microphylla* is very plentiful, and bushes 2 to 4 feet tall were covered with large pink blossoms. One small tree of *Carrieria calycina*, laden with curiously-shaped, waxy-white flowers borne in erect panicles, was also worthy of note. But the flora generally has been destroyed to make way for crops of maize, oats, and pulse.

Hei-shih ch'ang, alt. 4000 feet, is considered to be 60 li from Hsuan-kou, and consists of three or four houses, situated in a ravine alongside a torrent, with wild mountains on every side. Our lodgings were roomy, and the people both courteous and attentive.

Rain fell heavily next morning when we started out, but ceased about 9 a.m.; the weather remained dull the rest of the day until 4 p.m., when rain recommenced to fall and continued far into the following night. Crossing the torrent by means of a covered wood bridge the road immediately ascends a steep mountain called Che shan from the abundance of Varnish trees growing thereon. The ascent, though very steep, is short, and afterwards for the next 30 li the road skirts the mountain-sides until the summit of the Chiu-lung shan is reached. Descending this ridge it ultimately enters a narrow grassy valley. Here we found lodgings for the night in the solitary hostel of Hoa-tzu-ping, alt. 6100 feet, having covered 50 li during the day.

Until reaching the valley the country generally was either

under maize or covered with a dense jungle.   The flora was of passing interest only, being similar in character to that found everywhere in western Szechuan between 4000 and 6000 feet altitude.   The more interesting shrubs collected were a yellow-flowered Schisandra, a white-flowered Clematoclethra, and the Yunnan Holly (*Ilex yunnanensis*) with small, neat leaves, clusters of purplish, fragrant flowers, and hairy shoots. *Actinidia Kolomikta*, a large climber with white, fragrant flowers and added beauty in the shape of a multitude of white leaves, is excessively common.   Nearly all the species of Actinidia and the allied genus Clematoclethra, other than those clothed with rufous hairs, have these white leaves, which usually become pinkish as the season advances.   All the species are handsome climbers, and the majority bear very palatable juicy edible fruit.

The trees of this region, though not numerous or of any great size, include such remarkable subjects as Davidia, Ptero-styrax, Tapiscia, Tetracentron, Beech, and Horse Chestnut. Occasional trees of *Cornus kousa* occur, and were a wealth of white flower-heads enlivening the country-side.   Walnut trees are common around houses and wild strawberries by the wayside.   In the grassy valley the beautiful *Ilex Pernyi* occurs with *Rodgersia æsculifolia* and *Lilium giganteum* in quantity.   Around Hao-tzu-ping odd patches of maize are cultivated, but where clearings have been made the ground is mostly covered with grass and coarse herbs.

During the day we met many men laden with huge logs of Teih-sha (*Tsuga*, Hemlock Spruce) and Hung-sha (*Larix*, Larch) timber.   These logs were dressed, and carried on a wooden framework.   I measured one with a tape ; it was 18 feet 6 inches long, 7 inches thick, and 9 inches broad.   It is astounding how such loads are carried over vile mountain roads.   As fellow-travellers during the day we had some tribesmen in charge of a small mule caravan of tea, bound for the state of Wokje.

After leaving Hoa-tzu-ping we soon reached the head of the valley which merges into a narrow jungle-clad ravine. After a precipitous climb of 30 li we reached the summit of the Niu-tou shan, alt. 10,000 feet, where dense mists blotted out the landscape. A similarly precipitous descent of 20 li brought us to Chuan-ching-lou, where we put up for the night.

The flora was very interesting, but owing to a thick pall of mist I was able to observe only the plants immediately along-side the pathway. Perhaps the commonest shrub of the day was *Salix magnifica*, which is abundant everywhere, but more especially near the watercourses. This extraordinary Willow has leaves up to 8 inches long and 5 inches wide, with catkins 1 foot or more long. It forms a straggling bush 5 to 20 feet tall and, except when in flower or fruit, would scarcely be taken even by the closest observer for a Willow. (I first discovered this plant in 1903, and in 1908 succeeded in intro-ducing living plants into cultivation.) Many other kinds of Salix, varying from prostrate shrubs to small trees, occur on the Niu-tou shan ; indeed, this mountain is remarkable for its wealth in Willows (subsequently I succeeded in introducing into cultivation about a dozen species from this locality). The Actinidia and Clematoclethra previously noted again very abundant. *Clematis montana*, var. *grandiflora*, with large white flowers, was a pleasing sight ; so also was a Deutzia (*D. rubens*) with pretty rose-tinted flowers. I saw no deciduous broad-leaved trees of any size, but herbs were luxuriant every-where, especially the Rodgersia, which covers acres of the moun-tain-side. The Conifers were the most interesting plants of the day. In the ascent, save for odd trees of Silver Fir and Yew, I saw nothing but Hemlock Spruce. This tree delights in rocky country, clinging to the cliffs in a most remarkable manner. In the descent, however, Silver Fir, Spruce, Larch, Hemlock, and White Pine all occur, but the trees are being rapidly felled, and no large specimens were to be seen. From this place come the logs of timber noted yesterday. The Larch (*L. Mastersiana*) is first met with below T'ang-fang, alt. 9400 feet, where it is common more especially to the right of the road, and descends to 7200 feet altitude.

Chuan-ching-lou, alt. 7000 feet, 50 li from Hoa-tzu-ping, consists of one large, dirty hostel, and three other houses, situ-ated in a narrow ravine, walled in by lofty mountains. A noisy torrent which descends from the Niu-tou shan flows past the inn, and vegetation is rampant on all sides. The road over

the Niu-tou shan is difficult, and in many places dangerous.
Here and there steps have been cut in the hard rock to assist
the traveller, but in the main the road is strewn with loose
stones and boulders—vile to walk on or over.

We were unfortunate in the matter of weather, for it again
rained as we continued our journey. Following the torrent
through a narrow ravine for 5 li we reached Êrh-tao chiao,
where the torrent connects with a very considerable stream
which flows from the Pan-lan shan. The united waters form a
river which, after traversing very wild country, joins with the
Min near the foot of the Niangtsze-ling on the Wênch'uan
Hsien side of the pass. Turning sharply to the left at Êrh-
tao chiao we ascended the stream, which is called Pi-tao Ho,
and soon crossed over by a wooden semi-cantilever bridge to
the left bank. From this point the next 25 li to Wu-lung-kuan
is easy, going through a narrow valley where occasional houses
occur and a certain amount of cultivation obtains. Above
Wu-lung-kuan the road becomes increasingly difficult, and in
many places is execrable. The river is joined by numerous
lateral torrents, some of large size, and as the valley narrows
into a ravine becomes an untamable, roaring torrent. The
scenery, such as the mists permitted of our seeing, is savage
and grand. Here and there perpendicular cliffs of limestone
cropped out through the mists, their summits covered with
Pine trees. We crossed and re-crossed the torrent many times,
and after covering 65 li reached Ta-ngai-tung, which was our
destination for the day. This hamlet, alt. 7600 feet, consists
of one large hostel, which was in moderately good repair, and is
completely surrounded by steep mountains heavily clad with
mixed shrubs and small trees, the upper parts being covered
with forests of Conifers. The flora generally is very similar
to that of the Niu-tou shan, though scarcely as rich. All the
Conifers except Silver Fir are present, though Larch only puts
in its first appearance near the hostel. At Êrh-tao chiao
I photographed a magnificent Juniper tree, 75 feet tall, 22
feet in girth, with graceful pendent branches, and a Black
Pine which retains its cones over many years. (It proved to
be a new species, and has been named *Pinus Wilsonii*). This
Pine is common on the cliffs, but White Pine (*P. Armandi*)

is rare, although we passed the largest specimens of this tree I have ever met with. *Deutzia longifolia* with lovely rosy-lilac-coloured flowers, *Spiræa Henryi* with yard long, flat sprays of pure white, and *Neillia longeracemosa* with rose-coloured flowers were perhaps the commonest shrubs in blossom. Poplar is the only large deciduous tree hereabouts. Maple is not uncommon, and near Ta-ngai-tung I gathered specimens of a Black Birch having short, stout erect catkins.

Early next morning we continued our journey, spending the whole day toiling up the ravine through wild and savage, yet wondrous, scenery, with a profusion of vegetation on all sides. Coniferous trees preponderate, the species being the same as those previously mentioned, with a couple of new Spruces added. Yew is less abundant, but Larch much more so, though large trees are very scarce. To my astonishment the Larch cones were ripe, and I collected a quantity of seed. A Poplar with large leaves, silver-grey on the under side, is very common, and we passed some very large specimens. A Rose with large bright red flowers made a fine display, so also did the pink-flowered Deutzia mentioned above. Two Lady-slipper orchids (*Cypripedium Franchetii* and *C. luteum*), with rosy-purple and yellow flowers respectively, occur, but are rare. In the bed of the torrent *Hippophaë salicifolia* (Sallow-thorn) is common, and varies from dwarf spiny bushes to trees 25 feet tall, the long slender foliage silvery-grey below forming a pleasing contrast to the brighter greens of surrounding trees and shrubs. Many kinds of Maple (*Acer*), Linden (*Tilia*), and Mountain Ash (*Sorbus*) are plentiful, and *Tetracentron sinense*, an interesting tree exceeding in size all other deciduous trees of this particular region, occurs sparingly. Hydrangeas, Spiræas, Honeysuckles, Mock-orange, Brambles, Roses, Actinidia, Clematoclethra, Viburnum, and other ornamental shrubs struggle for possession of every available spot. The variety and wealth of bloom was truly astonishing, and I know of no region in Western China richer in woody plants than that traversed during the day's march.

The weather continued exasperatingly showery, but luckily no great quantity of rain fell, otherwise the route would have been impassable. Heavy mists limited our view, but whenever the clouds lifted we saw nothing but steep mountain-sides, beetling crags or cliffs, bare here and there but mostly clothed with mixed vegetation, giving place ultimately to forests of Conifers. The road is vile beyond the power of language to describe. In several places poles have been fixed horizontally into holes made in the face of the cliffs and half-rotten planks laid on these to form a roadway. Such bridges as exist are of logs, often rotten, and were always difficult to cross. The river is simply a roaring torrent, cascading over huge boulders and madly endeavouring to escape to less savage regions. At one point it receives a torrent, which, judging from the colour and temperature of the waters, evidently comes down from eternal snows.

During the day we passed a few miserable hovels, but there is no room for cultivation, and the people are wretchedly poor. We stayed for the night at Yü-yü-tien, alt. 8800 feet, 42 li from Ta-ngai-tung, where there are two poor hostels. These useful if squalid structures are all alike on this route, being one-storied, constructed of wood, and roofed with shingles held down by stones. A portion is sectioned off as private quarters for the family in charge, and near by the kitchen is located. A series of bunks is built around all sides of the place, the central part being occupied by benches for the accommodation of loads. Travellers furnish their own food-supplies, since nothing is obtainable at the hostel except, perhaps, some green vegetables in minute quantities. Shelter for the night and a fire to cook food and dry clothing are all these places afford. But the foreign traveller enjoys a welcome quietude and freedom from curious crowds. A sound night's sleep rewards the labours of the day, and he awakens refreshed, perfectly fit, and all eager to drink in more of the wondrous scenery, the charm of woodland, crag, and stream.

At Têng-shêng-t'ang, 8 li beyond Yü-yü-tien, the ravine widens out into a shallow valley, and the road boldly ascends the grassy, scrub-clad mountains to the left of the stream. Hereabouts Barberries in great variety luxuriate. After a severe ascent we crossed over a shoulder, and for the rest of the day skirted the side of a grassy ridge carpeted with brilliantly coloured alpine flowers.

PRIMULA VEITCHII

The main stream takes its rise in some snowclad peaks, of which we obtained a glimpse and a photograph, but a considerable tributary flows down from the Pan-lan shan Pass. The mountains to the right of this affluent, and also to the right of the main stream, are forested up to 11,500 feet altitude with Spruce, Silver Fir, and Larch. The bed of the valley is covered with bushes of Willow, Hippophæ, and Barberries. Up to 10,000 feet altitude *Cypripedium luteum* is not uncommon on humus-clad boulders and in the margins of woods.

The flora of the grassy ridge leading up to the Pan-lan shan Pass is strictly alpine in character, and the wealth of herbs was truly amazing. Most of the more vigorous growing had yellow flowers, and this colour predominated in consequence. Above 11,500 feet altitude, the gorgeous *Meconopsis integrifolia*, which has huge, globular, incurved, clear yellow flowers, covers miles of the mountain-side. Growing on plants from 2 to 2½ feet tall the myriads of flowers of this wonderful Poppy-wort presented a magnificent spectacle. Nowhere else have I beheld this plant in such luxuriant profusion. The Sikhim cow-slip (*Primula sikkimensis*), with deliciously fragrant pale yellow flowers, is rampant in moist places. Various kinds of Senecio, Trollius, Caltha, Pedicularis, and Corydalis added to the over-whelming display of yellow flowers. On boulders covered with grass and in moderately dry loamy places, *Primula Veitchii* was a pleasing sight with its bright rosy-pink flowers. All the moorland areas are covered so thickly with the Thibetan Lady-slipper Orchid (*Cypripedium tibeticum*) that it was impossible to step without treading on the huge dark red flowers reared on stems only a few inches tall. Yet the most fascinating herb of all was, perhaps, the extraordinary *Primula vincæflora*, with large, solitary, violet flowers, in shape strikingly resembling those of the common Periwinkle (*Vinca major*), produced on stalks 5 to 6 inches tall. This most unprimrose-like Primula is very abundant in grassy places. The variety of herbs is indeed legion, and the whole country-side was a feast of colour. Silence reigns in these lonely alpine regions, a silence so oppressive as to be almost felt and only broken on rare occasions by the song of some lark soaring skywards. We flushed an occasional Snow-partridge and saw one or two flocks of Snow-pigeons, but bird-life generally was extremely sparse. Save a few voles and mice we saw no animals, but Bharal and Wolves were said to occur here, the former in quantity.

After travelling 38 li we reached the hostel at Hsiang-yang-ping, alt. 11,650 feet, and remained there for the night. This place is part temple, part inn, and is kept by a priest, to whose clothing and person water was evidently a stranger. The medicine Pei-mu (*Fritillaria Roylei* and other species) is common in this region, and as fellow-guests for the night we had a number of people engaged in digging up the tiny white corms of this plant. Some Chinese traders also were there buying up this medicine at 60 cash per ounce. In Chengtu it is worth, wholesale, 400 cash per ounce, so their profit is a handsome one. Among the medicine-gatherers were several Wokje tribesfolk, about 5 feet 8 inches tall, sturdily built, with straight noses and fearless expression. Two of their women were with them, and had they been clean and decently dressed they would have been decidedly handsome and attractive. We enjoyed during the day a certain amount of sunshine, interrupted by occasional showers, but soon after our arrival at Hsiang-yang-ping it commenced raining in torrents, and continued to do so far into the night.

It ceased raining before daylight, to our great joy. Making an early start we toiled slowly over the dreaded Pan-lan shan, crossing the pass in a dense, driving, bitterly cold mist. The ascent is nowhere difficult, and none of us suffered seriously from the effects of the rarefied atmosphere, in spite of the evil reputation this pass has for mountain-sickness. The ridge is narrow, razor-backed, the summit being composed of sandstone, with marble embedded, piled up at an acute angle and devoid of vegetation. Snow, unmelted from the winter, lay in odd patches immediately below the pass, and on all sides there was much fresh snow. The dense mists prevented any extended view, but what little of the region was visible was bare and desolate. Two or three of the lovely Snowbird, *Grandala cœlicolor*, were flitting around the snowy patches, their intense blue plumage contrasting remarkably with the white carpet around. I made the pass, 14,250 feet altitude, and the tree-limit about 11,800 feet.

The flora above 12,000 feet altitude is purely alpine and similar in character to that of the region around Sungpan and elsewhere throughout the Chino-Thibetan Borderland at the same altitude. *Meconopsis integrifolia* occurs in countless thousands; also, to my pleasant surprise, the dark scarlet-flowered *M. punicea*. Although by no means so plentiful as around Sungpan, there were many thousands of this beautiful herb scattered around. Primroses are most abundant; *Primula vincæflora* ascends to 13,000 feet, where its place is taken by the lovely *P. nivalis* and another closely allied species.

On crossing over I photographed the pass and then descended with all possible speed to the miserable hostel of Wan-jên-fên, alt. 13,700 feet, where our lunch awaited us. A little below this hostel a few bushes of Willow, small-leaved Rhododendrons, and Caragana spp. first appeared and became abundant as we descended. Soon Larch and occasional Spruce appear, and at 11,300 feet altitude trees are fairly numerous. A shrubby, evergreen Prickly Oak is characteristic of these wind-swept mountain-sides, the golden-brown undersurface of its leaves rendering it most conspicuous. (This Oak is almost as beautiful as the Golden Chestnut of California (*Castanopsis chrysophylla*), and I am very pleased to report its successful introduction to cultivation.)

In addition to the shrubs mentioned above, dwarf Juniper, Spiræa, and Sallowthorn also abound. This moorland country is very interesting and shows unmistakable signs of a drier climate than that enjoyed by the regions on the opposite side of the pass.

A torrent which rises near the head of the pass is soon augmented by tributaries and quickly becomes a roaring unfordable stream. The mountain-slopes close in, and at the tiny hamlet of Kao-tien-tzu the road plunges into a ravine. The sides of this ravine are wooded, Larch and Spruce being abundant, with miscellaneous shrubby vegetation. The elegant *Syringa tomentella*, a Lilac with branching panicles of fragrant flowers, is very common. On issuing from this ravine we crossed a tributary torrent, more turbulent in character than even the main stream, and found in front of us open country largely under cultivation.

Our caravan was to have stopped for the night at the hamlet of Kao-tien-tzu, but with greater zeal than knowledge pushed on 20 li farther to Reh-lung-kuan. This blunder upset my plans and put all things awry. The collecting work had to be curtailed; it was 10 p.m. before I got any supper, and much of our work had to remain over until the morrow.

The Pan-lan shan is the boundary between two Chiarung states. On crossing over we quitted the state of Wassu and entered that of Wokje. The Wassu territory is wildly mountainous, well forested, and but little suited to agriculture. In consequence it is sparsely populated, and we encountered very few of the inhabitants *en route*. The hostels and houses on the main road are in the hands of Chinese or half-castes. The men of Wassu are tall (5 feet 8 inches or thereabouts), with large, muscular frames, frank, open countenances, and are noted hunters of the beasts of forest and crag. The women are sturdy, buxom, and engagingly frank. Both men and women are darker complexioned than the Chinese, and, I am sorry to say, infinitely less cleanly in appearance. They are very fond of jewellery, both sexes wearing bangles of silver and copper, and silver rings studded with coral and turquoise. The women also wear large silver ear-rings, usually having insets of coral and turquoise. The men are addicted to opium-smoking, though possibly this is strictly true only of those engaged near the main roads as porters and muleteers who have come in close contact with the Chinese.

Reh-lung-kuan, alt. 10,900 feet, is a Wokje village consisting of about a score of houses, a small lamasery, and a tall square tower. We found here a spacious and very fair inn, and the people were courteous and obliging. Our carrying coolies were able to purchase opium and a certain amount of food-stuffs. This explained their anxiety to cover 75 li, instead of stopping 20 li short at Kao-tien-tzu.

On leaving Reh-lung-kuan we descended the right bank of the river, which rises near the Pan-lan shan pass, for 33 li to the hamlet of Kuan-chin-pa, a short day's march being necessary in order to accomplish the work left over. The day was fine and warm, with a strong, cool breeze. Looking back on our route the snows of the Pan-lan shan were visible the whole day. The road was in good repair, and skirts the mountain-sides well above the stream. In ancient times this valley was filled with glacial detritus, through which the strong torrent has cut a deep, narrow bed. This stream, known locally as the Nei chu, is really the principal branch of the Hsaochin Ho (Little Gold River). Formerly gold in considerable quantities was mined in this valley, and we passed many old workings during the march.

The country generally reminded me forcibly of the Upper Min Valley, near Sungpan, above 8000 feet altitude. On the left bank of the stream the mountain-sides are very steep and largely covered with woods composed of Spruce, Silver Fir, and a few Pine trees. On the right bank the mountains are more sloping and mainly under cultivation. Wheat is the staple crop and ripens in early August; buckwheat ranks next in importance, followed at a respectable distance by peas, beans, and Irish potato. The Wokje people are evidently skilled agriculturists and in their own way fairly well-to-do. The prosperous condition of this state was evidenced by the plenitude of large houses, lamaseries, and by the relatively dense population. The hostels, however, are all in the hands of half-breeds, descended from early Chinese colonists. The larger houses and lamaseries are usually perched on some bluff composed of glacial mud, grits, and boulders. They are more or less square, two-storied, with flat mud roofs, having small turrets at each corner, from which prayer-flags flutter; a branch of some kind of Conifer is usually in evidence near these flags. Chortens and other Lamaist monuments occur here and there, while inscribed Mani-stones are common. The peasants' houses are low, one-storied, built of sandstone shales, the roof either flat or with very slight slope.

That the climate of this valley is relatively dry and warm is clearly shown by the flora, which is markedly xerophytic. Two species of Cotoneaster, several Clematis, the Sallowthorn, Prickly Oak, Barberries, and Roses are the chief constituents. A curious Bush Honeysuckle, with small leaves and tubular, white, fragrant flowers borne in pairs, is locally abundant.

(This proved to be a new species and has been named *Lonicera tubuliflora*.) Another common plant is the shrubby *Clematis fruticosa*, with simple oblong leaves and golden yellow, nodding flowers. A Lilac (*Syringa Potaninii*), with erect panicles of rose-purple flowers, is another interesting shrub, plentiful in this valley. Poplar, a Hard Pine (*Pinus prominens*), with almost prickly cones, and a White Birch, the bark of which is used for lining straw hats, are the more common trees by the wayside. I also gathered a few late flowers of *Incarvillea Wilsonii*. In a general way this Incarvillea resembles Delavay's species, but averages 4 to 6 feet in height. Another new plant collected was a Primrose akin to *Primula sibirica*, but with taller scapes and longer pedicels.

Kuan-chin-pa, alt. 9500 feet, consists of two small and rather poor inns, with the ruins of a large square tower near-by.

Twelve li below Kuan-chin-pa, and also on the right bank of the river, is the village of Ta-wei, a considerable place for this region, boasting a large lamasery. This place has an evil reputation, but no ill-will was displayed toward me. Many Lamas clad in claret-coloured serge crowded around and watched me as I photographed the village, and displayed much interest in my camera, dog, and gun. Nevertheless, the reputation of this village is well founded, and I would advise travellers to avoid staying overnight there. From Ta-wei a road leads across the river and over the mountains to Mupin.

On continuing our journey we followed the right bank of the stream for a further 27 li to Mo-ya-ch'a, where, owing to an old landslide, it was necessary to cross over to the left bank. This was accomplished by means of a wooden semi-cantilever bridge. Such bridges have been fairly common *en route*, but this was the first our road had led over. From this bridge the road descends the left bank, keeping high up above the river to Kuan-chai, which was our destination for the day. The whole valley is very arid, though a considerable area was under wheat. A few Poplar and Willow trees occur near the river, otherwise only high up on the mountain-sides were any trees discernible. The flora is similar to that of all the principal river-valleys of this hinterland, as described in Chapter XII. *Rosa Soulieana* is very abundant. I gathered several new plants, but the country is too arid to be of much interest botanically.

Situated at an altitude of 8500 feet, Kuan-chai is a small village and the residence of the Wokje chieftain. The chief's house is very large, the upper structure, all of wood, is well built, and the whole is dominated by several tall towers, and fine Walnut trees occur scattered around. The prosperous condition of this little state was further evidenced during the day's march. Large houses are frequent, many being perched high up on the steep mountain-sides. Wheat is the principal crop grown, and at Kuan-chai was just bursting into ear. Maize and the Irish potato are likewise commonly cultivated. A little flax and Hemp (*Cannabis*) also occur, the oil expressed from the seeds of these plants being in general use as an illuminant. We passed odd fields of opium poppy, the plants being only a few inches tall. On the fan-shaped slope, at the head of which the village of Kuan-chai is situated, all the crops were remarkably luxuriant.

At Ma-lun-chia a considerable torrent joins the Nei chu on the right bank. A by-road ascends this tributary, leading to Fupien and thence to Lifan Ting. Our road was for the greater part good and we easily covered the 67 li, enjoying bright sunshine the whole day.

Immediately beyond the chief's residence the road mounts over a steep bluff, where is situated the hamlet of Hsao-kuan-chai. This place is reputed to have offered a stern resistance to the Chinese in their conquest of this valley a hundred odd years ago, and was only captured after a long siege. The remains of the sangars and old forts are still to be seen. From this point the road continues to wind along the left bank of the river for 40 li to the town of Monkong Ting. Both sides of the valley are very arid, and the flora poor and uninteresting. Very few houses occur in the valley, but high up on the mountain-sides we saw many scattered about and surrounded by wheat fields. At Laoyang the river is joined on the right bank by another of almost equal volume. The main road from Lifan Ting, via Fupien, descends this tributary stream and joins at this point the road we were following. From what little we could see of the valley of this Fupien stream it appeared to be as arid and barren as the one we had descended from Reh-lung-kuan. Continuing our journey and on rounding a bend in the river, we suddenly sighted, perched on a rocky promontory, the town of Monkong Ting. After passing through a gateway we noticed a separate township, rather prosperous looking, situated in a lateral valley a little to the left of the main road. This is the official town of Monkong Ting, where reside the principal officials, civil and military. Crossing a torrent by a wooden bridge we entered the place first sighted from the bend in the river. This proved to be an old military camp of poverty-stricken, dilapidated houses, scattered alongside a street about 100 yards long. Two hundred yards beyond this camp we reached the thriving business town known as Hsin-kai-tsze. Monkong Ting, therefore, consists of three distinct towns or villages : (1) the official town, (2) an old military camp, (3) the business town. All three are unwalled, though a gateway has to be passed on entering each. The situation is most picturesque and strategically very strong. Monkong Ting is the political capital of this region and a place of very considerable importance. The two Chiarung states of Wokje and Mupin have their boundaries at this point, and the rest of the valley to Romi Chango is divided into feudal states.

The streets of Hsin-kai-tsze were thronged with people, chiefly tribesfolk, selling medicines and buying various articles for their own use. They made a very picturesque crowd, the women being especially noticeable by reason of their display of silver dress-ornaments, bangles, and ear-rings. The inns were all crowded, but the head official obligingly secured a couple of rooms for us and treated us with much courtesy and goodwill. The people were naturally curious and grouped themselves around us, but their manners were deferential.

Hsin-kai-tsze, alt. 8200 feet, is a most important medicine mart, being famous for its " Pei-mu " (*Fritillaria* spp.), " Rhubarb," " Ch'ung-tsao " (a caterpillar infested with the fungus *Cordyceps sinensis*), and "Chung-hoa" (an Umbelliferous plant, possibly *Ligusticum Thomsonii*). All of these are collected and brought in for sale by the tribesfolk. Musk and deer-horns also figure in the trade.

Several roads radiate from this centre ; one of these leads from the official town to Mupin, over the pass of Chia-chin shan, which was said to be higher than that of the Pan-lan shan and surrounded by snow-clad peaks.

The Wokje state preserved its prosperous appearance to the end, and is evidently a thriving, happy little country. The people strongly resemble the Wassu folk, though possibly they are scarcely as tall and have slightly sharper features. The Chinese language is understood and in common use along the main road, where the people imitate the Chinese in shaving their heads and wearing a queue. Lamaism evidently has a strong hold on these people, judging by the number of lama-series we saw.

I had intended remaining a day at Monkong Ting, but owing to the crowded condition of the town decided to defer this holiday until we reached Romi Chango. The inn in which rooms were provided for us was crowded with persons who were noisy over their cups and business dealings far into the night, rendering sleep well-nigh an impossibility.

Just outside Hsin-kai-tsze the road crosses over by a log bridge to the right bank of the stream. This bridge was being repaired, and only two very uneven logs were in position. A thin rope was stretched across to serve as a hand-rail on the left side. Crossing was really dangerous, the waters below being deep and turbulent. The official kindly provided local experts to carry our gear over, and the way these men accomplished the task filled me with admiration. I rewarded them with 1000 cash, to their astonished delight. My dog was lashed firmly to a flat board and carried across on a man's back. He struggled violently, and the man only just managed to get him over before he got half loose. I walked over behind the dog and was relieved when the 30 yards across the yawning gulf were safely passed. Everything came over all right, but my followers clung to the local men like grim death, the majority shaking in their nervous fright. Such dangerous experiences are not desirable, and I heartily hoped that we had no more such bridges to cross. From this bridge we descended 60 li to the hamlet of Shêng-ko-chung, alt. 7600 feet, through arid country and over a bad road. The river is here a broad and turbulent stream, flowing between steep banks composed of loose rocks. A few Poplar, an occasional Cypress (*C. torulosa*), and the Kœlreuteria (the latter was covered with masses of small yellow flowers) are the only trees of note. The region is very sparsely populated, but high up, on the left bank more especially, are a few houses of the same architecture as those of Wokje.

As travelling companions during the day we had a party of tribesfolk, chiefly women in holiday attire. They were very cheerful, laughing and singing most of the time. On parting company at Shêng-ko-chung they made merry over cups of Chinese wine, the dames officiating as to the manner born.

It rained heavily during the night, and it was cool and delightfully fresh in the morning when we recommenced our journey down the valley of the Hsaochin Ho. Thirty li below Shêng-ko-chung we passed the large lamasery of Gi-lung, coloured white and picturesquely situated on the right bank of the river. Over a hundred Lamas reside here and exercise considerable authority over the neighbourhood. About 10 li beyond this lamasery the river suddenly develops into a series of boiling, roaring cataracts. The fury of the waters was most fearsome to behold, and a wilder stretch of river is scarcely imaginable. Earlier in the day we had crossed to the left bank, and just below the very worst bit of this savage waterway we recrossed to the right bank over a rotten and most unsafe wooden bridge. Some 7 li below this point we reached the hamlet of Pan-ku chiao, alt. 7100 feet, where we found accommodation for the night, having covered 70 li. Just above the hamlet a torrent joins the river on its left bank, and up this lateral valley mountains clad with snow were plainly visible. Bridges are scarce and the few that exist look as if they had not been renewed since this region was conquered, well over a hundred years ago. One thing is certain, they cannot possibly last much longer : the two we crossed during the day were all askew and decidedly dangerous.

The district is rather less arid than that around Monkong Ting, yet the flora is very poor. Poplar is a common tree, so also is the Kœlreuteria, which was a fine sight, with a wealth of flowers, and it evidently enjoys a dry, hot situation. The

THE GI-LUNG LAMASERY

sub-shrubby *Incarvillea variabilis* and *Amphicome arguta*, both with large, tubular, pink flowers, are very abundant by the roadside. Other common shrubs are Bauhinia, *Sophora viciifolia*, Ceratostigma, with lovely blue flowers, Ligustrum and *Rosa Soulieana*. On the cliffs *Cupressus torulosa* is dotted about. Maize is the principal crop, occupying in season almost every inch of available land. Houses are fairly numerous, but most of them are relegated to the higher slopes well above the valley. The scenery in places is rugged and grand. In front of the inn at Pan-ku chiao limestone cliffs rear themselves some 2000 feet, abutting on a cultivated slope where Walnut trees are scattered around. Crowning a bluff is a tall tower, and near-by another in ruins, telling of glories now departed.

On leaving Pan-ku chiao we descended the right bank of the Hsaochin Ho, some 42 li to the point where it joins the Tachin Ho or Upper Tung River. This final stretch is little else but one long succession of cataracts and strong rapids, the turbulent waters being thick with brown mud. High bare cliffs predominate, but here and there occur more or less flat fan-shaped areas under cultivation, with houses shaded by Poplar, Willow, and Walnut trees. *Diospyros Lotus, Hovenia dulcis*, and the large-leaved *Ligustrum lucidum* are other trees common hereabouts. Maize is evidently the chief summer crop in these regions, but wheat is grown, a red, beardless variety, with stout ears, and harvesting was in progress. Rock-pigeons are very abundant, and were busily engaged in exacting their toll of the ripening grain.

After passing the hamlet of Yo-tsa we sighted on the opposite (left) bank a large lamasery sequestered midst a fine grove of trees. A little beyond this is the village of Tsung-lu, a curious-looking place, boasting of a score or more tall towers. Skin coracles are employed to ferry over to these places.

The Hsaochin Ho is prevented from joining the Tachin Ho at right angles by a rocky spit which at times is evidently flooded over. Marble and granite are common rocks hereabouts, the latter being full of mica flakes which glistened in the sun. Ascending the left bank of the Tachin Ho for a couple of li, then crossing over a bamboo suspension bridge 90 yards long, we soon reached the small town of Romi Chango. The whole day's journey was only 45 li, but owing to the heat and rough road we all arrived very much fatigued and in sore need of a day's rest.

From all I could learn it would appear that the region in the vicinity of the river from Monkong Ting to Romi Chango, after its conquest by the Chinese about A.D. 1775, was divided into feudal states, and certain chieftains installed in possession as rewards for services rendered during the struggle. The chiefs, styled Shao-pês, hold hereditary office and are directly responsible to Chinese authority for the good behaviour of the people under their rule, also, if necessity arises, they are bound to supply armed men to assist the Chinese cause. Lamas alone are exempt from such military duties ; ordinarily the people of these feudal states are agriculturists. These Shao-pês are subordinate to the Chinese military commander stationed at Monkong Ting. The two chief Shao-pês reside, one at Monkong Ting, the other at Che-lung, a village in the mountains, 20 li removed from the left bank of the Hsaochin Ho and 60 li below Monkong Ting. Another Shao-pê resides at Ta-ching, 120 li to the north-east of Monkong Ting ; a fourth at A'n-niu, a place in the mountains to the south-west of the region controlled by the Che-lung Shao-pê. Beyond the original grant of territory these feudal chiefs receive no rewards, monetary or otherwise, from the Chinese. The system has much to recommend it and evidently works very well. It keeps the Chinese authority supreme, while it allows the native people to be governed by their own recognized chiefs. The difference between the chieftain of a semi-independent Chiarung state and a Shao-pê appears to be that, whereas the former is an absolute ruler over a territory long hereditary to his tribe, the latter is more in the nature of an alien ruling over a tract of country fiefed to his forbears by the Chinese, after they conquered this region and broke up the Chiarung confederacy. The territory occupied by these feudal states formerly belonged to the Chiarung tribes, and the people are principally derived from that stock. Chinese settlers have intermarried with the natives, and in the vicinity of the main road the population is mixed. The people living in the lower stretches of the Hsaochin Ho are an inferior race, of poor physique, and most abominably filthy.

CHAPTER XV

ACROSS THE CHINO-THIBETAN BORDERLAND

Romi Chango to Tachienlu ; the Forests of the Ta-p'ao Shan

ROMI CHANGO, or Chango, as it is commonly called, is a poor, unwalled, straggling town of about 130 houses. It is without rank, but a magistrate, subordinate to the Tachienlu Fu, and a military official, controlled from Monkong Ting, reside there. The town is really a Chinese settlement, situated in the extreme north-east corner of the state of Chiala. It is built on the right bank of the Tachin Ho, at a point where the river, making a right-angled turn from the northward, is joined by a very considerable torrent from the west. The Tachin, a river 100 yards broad, with a steady current and muddy water, sweeps round majestically. High cliffs on the left bank, steep mountain-slopes on the right, lofty mountains to the east and west wall in the town, at the western entrance to which a massive square tower stands sentinel. Chango is a very poverty-stricken place, with a small trade in medicines and sundries. It draws its supplies of rice, paper, and Chinese commodities generally from Kuan Hsien, and everything is phenomenally dear. This is only natural when the distance and difficulties of the journey are duly considered.

A small road descends the right bank of the Tachin Ho, by means of which Luting chiao may, with great difficulty, be reached. A road ascends the right bank of the Tachin Ho and leads to the interesting Chiarung states of Badi and Bawang, where the Bönpa religion holds full sway. Badi, the capital of these now united principalities, is only 60 li from Romi Chango. The chieftain is dead, but his widow, assisted by a steward, acts as regent for her infant son. Badi-Bawang is one of the ancient matriarchal kingdoms of Chinese historians, and at all times a woman holds an important place in its government. Badi, the larger of the two states, is very rich in gold, which, though unworked during recent years, is jealously guarded. Chinese visitors, rich or poor, are cross-questioned as to their business and closely watched during their sojourn in this state. The Badi-Bawang folk often visit Chango on business, and during our stay there we saw several. Most of them were peasant girls and women, dressed so scantily as to scarcely hide their nakedness. They were short in stature, and apparently unwashed from birth! However, since these were " hewers of wood and drawers of water " of the poorest class, it would be unfair to judge the whole race by them.

In Chango we lodged at a comfortable inn, having a clean room, well removed from the street and overlooking the river. We spent a quiet day resting and refitting for the final stage of our journey to Tachienlu. The people were not over-inquisitive and those in charge of the inn were exceedingly obliging. Soon after our arrival the magistrate sent me word that he was suffering from pains in the stomach and vomiting, and would be grateful for some medicine to relieve his suffering. I sent him some Epsom-salt and an opiate. The next day word came that he was much better, only too tired to leave his room. A traveller gets many such requests for medicine, and I have generally found quinine, Epsom-salt, and opium pills most useful cures, for which the people were always grateful.

On leaving this lonely town of Chango, which I made 6700 feet altitude, the road to Tachienlu ascends the right bank of the tributary torrent. We were warned that the road was very difficult, leading through forests and over high mountains. It was not long before these statements were verified. The torrent quickly develops into an angry, irresponsible stream ; the road in many places had been washed away and much wading was necessary. Our carriers had great difficulty in getting along, and had the waters of the torrent been a few feet higher the road would have been quite impassable. All the bridges were rotten and most insecure. High up on the mountain-sides we saw several large hamlets, but there are very

THE TOWN OF ROMI CHANGO

few houses in the valley—quite sufficient, however, for whenever the road led past a house we had to traverse an open
sewer, often a foot deep in dung and refuse. Such filthy
surroundings are characteristic of Thibetan houses. The
Chinese would collect all this sewage for their fields, but the
Thibetans, who are but poor agriculturists at best, have not
yet learned the value of manure. At such places I usually
climbed over the fences and walked through the crops, but my
men waded through the filth and gave vent to their wrath in
loud, angry imprecations. The people of Chiala are typical
Thibetans and use the lower stories of their flat-roofed houses
as pens for horses and cattle. A few li above Chango the flora
begins to lose its purely xerophytic character, and becomes
more and more luxuriant as the ascent proceeds. The higher
slopes are well forested with mixed trees, but near-by the road
trees are scarce. The mountain-sides flanking the stream
are very steep, being often sheer cliffs. Such places are
dotted with Cypress (*Cupressus torulosa*) and prickly leaved
Evergreen Oak.

After journeying 60 li we reached the village of Tung-ku,
alt. 7800 feet, where there are several large Thibetan houses,
decorated with prayer-flags, but only two or three hostels, and
these very poor in character. The owner of the one we stayed
in is a noted hunter, and many pelts of the Budorcas, Serow,
and Black Bear were in use as bed-mattresses. His family
told us the hunter was away after Musk-deer ; they also
informed us that both the Thibetan-eared and Lady Amherst
Pheasants are common hereabouts. Around the village
Walnut trees are most abundant. Wheat is a common crop
and was just ripening. Maize too was plentiful and is evidently
the staple summer crop everywhere in these regions.

The next day we covered another 60 li, putting up for the
night at the poor hamlet of T'ung-lu-fang. We crossed the
river four times by wooden bridges, each more rotten than the
other. The river was in partial flood, and a goodly portion of
the road was either washed away, obliterated by landslides,
or under water. Often we had to make a path for ourselves
up the mountain-side. The under-water portions of the road
I traversed on the back of one of the soldiers we had with us

from Chango, until he stumbled and gave me a ducking. After this I waded. There was no traffic on the road so called, and I marvelled how my coolies managed to get their loads along. Our chairs were carried piecemeal and even then with difficulty over the worst places. The river was a roaring torrent throughout the whole day's journey, in places really awesome to behold, dashing itself headlong over enormous boulders, or boiling as if forced up by some malignant spirit. In many places our path actually overhung this torrent, and one false step meant death.

About 10 li above Tung-ku the river makes a right-angled turn and is joined at this point by another stream of almost equal volume from the westward. From this place the road skirts the river through a narrow, savage, magnificently wooded ravine. Maple, Ash, Hornbeam, Birch, Poplar, Hemlock Spruce, and Prickly Oak are the chief constituents of these woods, followed by Evodia, Rhus, Cypress, Willow, Elm, Sallowthorn, Bamboo, and miscellaneous shrubs. The Maples (*Acer Davidii* and *A. pictum*, var. *parviflorum*) are larger trees than I have seen elsewhere. The Ash and Hornbeam are all fine trees, and the Hemlock Spruce in many cases over 100 feet tall, with a girth of 12 to 15 feet.

On leaving this magnificent fragment of virgin forest the country became less interesting. Where the cliffs are not sheer and bare the mountain-slopes have been cleared to a very large extent. The ravine widens into a narrow valley which is covered with scrub. The cliffs and mountain-slopes high up are sparsely clad with Cypress, White and Hard Pine, Spruce, Silver Fir, and Hemlock. The scenery is sublime.

We passed very few houses and these of the meanest description. Very little land is under cultivation ; Maize is the chief crop, with patches of wheat and oats here and there. The country is not suited to cultivation, and one marvels how the few people living there manage to find even the most miserable subsistence. Yesterday we noticed herds of a small breed of cattle. The people are shorter in stature than the average, and perfectly proportioned dwarfs are fairly common. Since leaving Monkong Ting, goitre has been manifest among the inhabitants, and in this river-valley it is very prevalent.

FORESTS WEST OF CHANGO, A POPLAR TREE IN FOREGROUND

T'ung-lu-fang, alt. 8800 feet, consists of about half a dozen scattered houses. The one we stayed in is of Thibetan architecture, fairly clean, and owned by a Chinese settler. None of these houses affords any bedding for the coolies, and of course nothing is purchasable—all food-stuffs have to be carried by the travellers themselves.

The people at T'ung-lu-fang informed us that we should not be able to reach Mao-niu, as the road had been badly washed away in several places, and under the lee of some cliffs was flooded to a depth of 4 feet or more. This gratuitous and discouraging information proved, luckily for us, to be scarcely accurate, since, after a struggle, we managed to get through. My head coolie declared it was the very worst road we had ever traversed, and I was inclined to agree with him. Worse it could not have been and constitute a roadway at all! For fully half the distance the track was under water or washed completely away, and we were forced to wade or make a new path over the mountain-side. Just how we got over the 30 li I cannot describe, but we all came through with nothing worse than a severe wetting.

Mao-niu is a fair-sized village for the country, and is mainly perched on a flat 200 feet above the torrent, and surrounded by a considerable area under wheat—a veritable oasis, in fact, surrounded by high mountains. Formerly it was the principal village of a petty state to which it gave its name. It now belongs to the state of Chiala. As far as Mao-niu the scenery and flora is similar to that around Tung-ku and calls for no special remark. The outstanding feature is the woods of Hard Pine (*Pinus prominens*). The steeper the country the happier this Pine appeared to be. The bark of the trunk is deeply furrowed, often red in the upper parts of the tree; the cones are quite prickly, and are retained for many years. The wood is very resinous, and is evidently much esteemed for building purposes. The Hemlock Spruce is common, and all the trees are of great size.

At Mao-niu the main stream leads off in a westerly direction to Th'ai-ling, a large village of over 100 houses and several lamaseries. It is also the centre of a considerable gold-mining industry, and has the reputation of being a lawless district.

We were informed that the road thither was in a dreadful state of disrepair, and that most of the bridges had been washed away by recent floods.

On clearing the cultivated area around Mao-niu we plunged immediately into a narrow, heavily forested ravine, down which a considerable torrent thundered. Conifers preponderate in these forests, Spruce being particularly abundant. We noticed some huge trees, but the average was about 80 to 100 feet tall. White and Red Birch are common, and I was fortunate enough to secure seeds of the latter. The Sallowthorn (*Hippophæ salicifolia*) is exceedingly common, forming trees 30 to 50 feet tall with a girth of 4 to 10 feet. The size of these trees very much surprised me. Willows, Cherries, and different species of Pyrus are also plentiful. Deutzia, Hydrangea, Philadelphus, Rosa, and Clematis are the principal shrubs, and many were in flower. *Primula Cockburniana*, which has orange-scarlet flowers, is the most noteworthy herb hereabouts.

After wandering several miles through the forests we reached the hamlet of Kuei-yung, alt. 10,100 feet, and 60 li from T'ung-lu-fang. This place consists of half a dozen houses, purely Thibetan in character, built on a slope and surrounded by a considerable area under wheat, barley, and oats. The mountains all around are heavily forested with coniferous trees, and in the far distance a snow-capped peak glittered on the horizon.

The house we lodged in is three-storied with the usual flat mud roof. The walls built of shale-rock are most substantial. Entering through a low doorway we had first to traverse a yard filled with cattle dung, then a piggery where a steep ladder led upwards to a couple of dark empty rooms in which we installed ourselves. A ladder from these rooms led to the roof, where I should have preferred to sleep had it not been raining. The house boasts neither table, stool, nor chair, and we had to improvise as best we could. The Thibetans squat on the floor for their meals, and therefore have no use for tables or chairs. The housewife, a most cheery if dirty person, had a very musical laugh. Things generally appeared a joke to her, and incited her to frequent laughter, which it was pleasant to hear. My followers were oddly amused at the strangeness of things, and appeared to enjoy the novelty.

Yet it was not out of love for our quarters that I stayed over a day at Kuei-yung, but to photograph various trees and investigate the Conifers. Photography in the forests is no mere pastime. It took over an hour on three occasions clearing away brushwood and branches so as to admit of a clear view of the trunk of the subject. I secured a dozen photographs, which entailed a hard day's work. The trees of Larch and other Conifers, Birch, and Poplar are very fine. The Larch (*L. Potaninii*), though not plentiful, is of great size, and trees 100 feet by 12 feet in girth occur. But the most astonishing feature of these forests is the large trees of Sallowthorn (*Hippophæ salicifolia*). I had never imagined it could attain to the size of specimens I saw during the day. I photographed two old trees 50 feet tall, 12 and 15 feet in girth respectively. I saw others taller but less in girth. Another interesting tree hereabouts is a Cherry (*Prunus serrula*, var. *tibetica*), which has a short, very thick trunk, and wide-spreading head. The leaves are willow-like, 3 to 4 inches long; the fruit is red, ovoid, on pendulous stalks. The tree averages about 30 feet in height, the head being 60 feet and more through.

The next morning we bade farewell to our cheery hostess at Kuei-yung, and continued our journey. The road immediately plunges into the forest, and winds through and among magnificent timber. The forests are very fine, and coniferous trees 100 to 150 feet tall, with a girth of 12 to 18 feet, are quite common. The latter consist of four species of Spruce, three of Silver Fir and one of Larch. The handsomest of the Silver Firs is *Abies squamata*, which has purplish-brown bark, exfoliating like the bark of the River Birch. The Larch becomes general in the ascent, and ultimately overtops all other trees and extends to the tree-limit. White and Red Birch, Poplar and Sallowthorn are the only broad-leaved deciduous trees really common. An Evergreen Oak (*Quercus Ilex*, var. *rufescens*), with prickly leaves like a Holly, is abundant. In the shelter of the forests this Oak makes a good-sized tree, but in the more exposed places it is reduced to a small shrub. The wood is very hard and makes the finest of charcoal. Shrubs are not rich in variety, but Bush Honeysuckles, Barberries, Spiræas, and Clematis are plentiful. Herbs, especially the Sikhim cowslip (*Primula sikkimensis*), *P. involucrata*, Anemone, Caltha, Trollius, and various *Compositæ* luxuriate on all sides, and the glades and marshy places were nothing but masses of colour. The men who were in front of me saw several troupes of monkeys and some Eared-pheasants, but I saw no animals and very few birds.

We camped near the tree-limit, at about 12,000 feet altitude, and erected a small hut of spruce boughs under a large Silver Fir tree. My Boy preferred to pass the night in his chair, and the men arranged themselves around a log fire. The neighbourhood has an evil reputation for highway robbers, but we felt sure there was small possibility of any attack on us being made. It rained a little during the day, and a sharp shower fell in the early evening, but the night proved fine. The altitude, however, affected our sleep ; it was also very cold, and we were all glad when morning broke. My dog suffered as much as any of us ; he refused to eat his supper, and I never saw him so utterly miserable. The coolies looked a most woebegone crowd, shivering with cold and generally wretched. They seemed to have no idea of making themselves comfortable ; it would have been a simple matter for them to have rigged up a shelter of spruce boughs, but they were too indifferent to do this or even to collect firewood. We brought with us from Kuei-yung, as guide, a Thibetan, and it was he who got together all the wood required for a fire.

There was a slight frost and a heavy dew, but the sun, which rose like a ball of fire, soon warmed us and dispersed the dew. The road is of the easiest, winding through timber and brush alongside a small stream, up to within 1000 yards of the head of the Ta-p'ao shan Pass, where the ascent becomes steeper. It is, however, only the last 500 feet that make any pretence of being difficult. Above the place where we camped the Conifer trees rapidly decrease in size, Larch becomes more and more abundant, and ultimately forms pure woods. It overtops every other kind of tree, and extends up to 13,500 feet altitude. Just below the limits of the Larch a dwarf Juniper appears and ascends to near the head of the pass. The scaly-barked Silver Fir (*Abies squamata*) ascends to 12,500 feet and two species of Spruce to 13,000 feet. This

RHEUM ALEXANDRAE AND OTHER ALPINE FLOWERS

side of the pass enjoys a moist climate, and the tree-line
(13,500 feet approximately) is remarkably high. Above the
tree-line the mountain-sides, to within a few hundred feet
of the pass, are covered with scrub composed, as usual in
these regions, of Willow, Berberis, small-leaved species of
Rhododendron,Spiræa, Juniper, *Potentilla Veitchii, P. fruticosa*,
and *Rhododendron Przewalskii*, the latter being the most
alpine of all the large-leaved members of its family. Herbs,
of course, made a wonderful display of colour. In addition
to those previously mentioned, other species of Primula, the
yellow and violet-blue Poppyworts (*Meconopsis integrifolia*
and *M. Henrici*), various Stone-crops (*Sedum* spp.), and Saxifrages are abundant. But the most striking of all the herbs
is a Rhubarb (*Rheum Alexandræ*), an extraordinary plant,
with a pyramidal inflorescence 3 to 4 feet tall, arising from
a mass of relatively small, ovate, shining, sorrel-like leaves,
and composed of broad, rounded, decurved, pale yellow bracts
overlapping one another like tiles on a house-roof. The local
name of this plant is " Ma Huang " (Horse Rhubarb) ; it
prefers rich boggy ground where verdure is luxuriant and
yak delight to feed. Such places were studded with its most
conspicuous tower-like spikes of flowers. The Rhubarb and
yellow Poppywort (*Meconopsis integrifolia*) are always most
rampant around places where yak have been herded.

Unmelted snow of the preceding winter was lying in
patches just below the summit of the pass, a bare, narrow ridge
crowned by a cairn of stones surmounted with many prayer-
flags, and 14,600 feet above sea-level. This narrow neck is
composed of slate and sandstone, with a certain amount of
marble rock scattered about, and connects two massive ranges
clad with eternal snows. The day was gloriously sunny, and
we had a rare opportunity of enjoying and appreciating the
delights of this alpine region. Except for a feeling of giddi-
ness when stooping, and a general shortness of breath, I
suffered no inconvenience from the altitude. In spite of their
loads only two or three of my men were seriously affected ;
the gradual ascent was, I think, responsible for our good
fortune in this matter. From past experience I had rather
dreaded the effects this pass might have on my followers, and

was pleasantly surprised at the ease with which they negoti-
ated it.

With the weather conditions so favourable the view from
the summit of the pass far surpassed my wildest dreams.   It
greatly exceeded anything of its kind that I have seen, and
would require a far abler pen than mine to describe it adequately.   Straight before us, but a little to the right of our
viewpoint, was an enormous mass of dazzling eternal snow,
supposed to be, and I can well believe it, over 22,000 feet high.
Beneath the snow and attendant glaciers was a sinister-looking
mass of boulders and screes.   In the far distance were visible
the enormous masses of perpetual snow around Tachienlu.
In the near distance, to the west-north-west of the pass,
another block of eternal snow reared itself.   Looking back
on the route we had traversed we saw that the narrow valley
is flanked by steep ranges, the highest peaks clad with snow,
but in the main, though bare and savage-looking, they scarcely
attain to the snowline.   On all sides the scenery is wild,
rugged, and severely alpine.   A cold wind blew in strong
gusts across the pass, and we were glad when our photographic
work was finished, and we could hurry down.   Several fine
Eagles and Lammergeiers were soaring aloft, but we saw no
animals, though Wild sheep and Thibetan gazelle were said to
frequent this region.

Descending by a precipitous, break-neck path, over loose
slate, sandstone shales and greasy clayey-marls for 15 li, we
reached the head of a broad valley.   The pass on this side
offers a far more severe climb than the side we had ascended.
On reaching the valley the track we followed connects with
the main road to Th'ai-ling, Chantui, and Chamdo.   Commercially speaking this is the highway into Thibet from
Tachienlu.   It leads through grasslands, affording good pasturage for animals, and though the mean elevation is very considerable the passes are less steep than those on the political
highway via Litang and Batang.   This Ta-p'ao shan region is
notorious for its highway robberies.   We met five tribesmen
who told us that in the previous night their camp had been
rushed by an armed band and everything they possessed
carried off.   Every Thibetan is by nature a robber, and

behaves as such when he fancies he can do so with impunity. They rob one another freely, but the tribesmen are their favourite victims.

From the head of the valley to Hsin-tientsze, the first habitation, is reckoned as 30 li. The road is broad but uneven, winding through a valley, and keeping close to a torrent which descends from the Ta-p'ao shan snows. The mountains on either side of the valley in all their higher parts range above the snowline ; their lower slopes are covered with grass, small Conifer trees, and brushwood. In the valley itself shrubs of large size, chiefly Willows, Honeysuckles, Barberries, and Sallowthorn abound. Odd trees of Larch and Spruce occur, all of small size. Flocks of Snow-pigeons were plentiful, and I shot several of these birds for our larder.

From Kuei-yung, 120 li, there is no house of any description save Hsin-tientsze, alt. 10,800 feet, a filthy and miserable hostel. Near Kuei-yung we passed a charcoal-burning camp where a few men were engaged, otherwise we did not meet a living soul, until we had crossed the pass. It is indeed a most lonely region, but of great interest to a Nature lover. I count myself particularly fortunate in being favoured by perfect weather for crossing the pass, more especially as it was the first day without any sign of rain since leaving Kuan Hsien.

The thermometer registered 36° F. when we turned out next morning, and our ears and fingers tingled with cold, even though it was 8th July. The smoke inside the inn was too much for my eyes, so I breakfasted out in the middle of the roadway. I think everybody was glad to quit Hsin-tientsze with its vermin and stinks. There was an odd patch of wheat around the hostel, but it looked miserable ; the season is too short and the climate too severe for cultivation hereabouts at this altitude.

We followed a broad, uneven road, which had suffered much from animal traffic, for 60 li to Jê-shui-t'ang (Hot-water pond), alt. 9800 feet. The descent is gradual, and the day's journey proved a delightful loiter through a shrub-clad valley. We met several hundreds of yak and ponies, all laden with brick tea encased in raw hides and bound for interior Thibet. The Thibetans in charge were an unkempt, wild-looking lot of men, with long guns, swords, and conspicuous charm-boxes. Many of them wore their hair in a long plait with a sort of black yarn braided in, the whole being wrapt around their heads to form a turban ; a few wore felt hats with high conical crowns. One or two women were with these caravans tending the animals exactly in the same way as the men. Ability to whistle and heave rocks with sure aim seemed to be the essential parts of a yak-muleteer's profession.　Yak are slow, phlegmatic animals, and on sighting any unusual object they stand stock-still for a little time, and then make a mad rush forward.　They appeared to be docile enough, but their long horns looked dangerously ugly, and we got out of their way as much as was possible.　Each caravan was accompanied by one or more large dogs.　These animals trot alongside the caravan and take no notice of any one, but when tethered and on guard in camp will allow no stranger to approach.　They are massively-built dogs, and their savage appearance is heightened by a huge red-coloured collar of woollen fringe, with which they are commonly decorated.

The flora was merely a repetition of that of the previous afternoon's journey.　The valley and contiguous hill-sides are covered with scrub, except for clearings here and there which serve as yak-camps.　In addition to the shrubs mentioned as occurring around Hsin-tientsze, Prickly Oak, Juniper, several kinds of Rose, and the Thibetan Honeysuckle (*Lonicera thibetica*) are common ; Barberries in variety are a special feature.　Conifers are scarce and all of small size ; all the larger timber has been felled and removed long ago　At the hamlet of Lung-pu, reckoned 40 li from Hsin-tientsze, crops of wheat, barley, oats, and peas put in an appearance, and became more general as we descended the valley.　Around Jê-shui-t'ang the cereals were just coming into ear.

During the day, which was beautifully fine, we had grand views of the snowclad peaks around Tachienlu and the steep ranges with pinnacled peaks to the east-south-east of that town.　Around Jê-shui-t'ang there are several hot springs, in some of which the water was actually boiling.　These springs are rich in iron, but in those I examined no sulphur was evident.

Our quarters at Jê-shui-t'ang were a considerable improve-

THE YA-CHIA-K'AN, PEAKS 21,000 FT.

menton those of Hsin-tientsze, but we were, nevertheless, glad to leave very soon after day dawned. It is considered to be 90 li from this place to Tachienlu, but I should say 60 li is a nearer estimate. We enjoyed another sunny day. The road is easy and leads through a continuation of the valley that we entered on descending from the Ta-p'ao shan Pass. The valley and mountain-sides for some 300 to 500 feet above it become more and more under cultivation. Cereals, peas, and Irish potato are the principal crops. The potatoes were being harvested, and I noticed that red ones predominated. The region generally has been denuded of its trees, and where not under crops is covered with scrub and coarse herbs. In rocky places small trees of White and Hard Pine (*Pinus Armandi, P. prominens*) occur, also a few comparatively large trees of a very distinct-looking Peach having narrow, lance-shaped, long pointed leaves, rather small fruits, downy on the outside.[1]

Around habitations tall trees of Poplar are common, and an occasional Spruce and White Birch occur. The Spruce (*Picea aurantiaca*) is a particularly handsome species, with square, dark green needles on spreading branches and red-brown pendulous cones clustered near the top of the tree. The Apple, Apricot, Peach, Plum, and a few Walnut trees are cultivated. The fields are fenced with hedges of Wild Gooseberry (*Ribes alpestre*, var. *giganteum*) and the handsome *Sorbaria arborea*, which has large erect masses of snow-white flowers. Over these and other shrubs various species of Clematis trail, the most common being *C. nutans*, var. *thyrsoidea*, which was laden with a multitude of creamy-yellow nodding flowers. The most beautiful shrub, however, was a Lilac growing 12 to 15 feet tall, and covered with huge panicles of pink or white fragrant flowers. (It proved a new species, and has been named *Syringa Wilsonii*.)

---

[1] At the time I paid no further attention to this Peach, but in 1910 I secured ripe fruit, and found to my astonishment that the stones were perfectly smooth, free, and relatively very small—characters denoting a distinct species of Peach. It proved to be new, and has since been named *Prunus mira*. I regard this as among the most remarkable of the discoveries I have been privileged to make. This new Peach is now in cultivation, and by crossbreeding with the old varieties of the garden Peach (*P. Persica*) may result in the production of entirely new and improved races of this favourite fruit.

We crossed the stream by a wooden cantilever bridge and, on rounding a bend, the goal of our long journey came into view. We were all well-nigh dead beat, and it was with thankful and joyous hearts that we greeted the cluster of closely packed houses, which, nestling in a narrow valley, constitute the important border town of Tachienlu.

CHAPTER XVI

TACHIENLU, THE GATE OF THIBET

The Kingdom of Chiala, its People, their Manners and Customs

THE town of Tachienlu is situated in long. 102° 13' E., lat. 30° 3' N. *circa*, at an altitude of about 8400 feet. By the most direct route it is twelve days' journey from Chengtu Fu, the provincial capital, on the great highway which extends westwards to Lhassa. It is the Ultima Thule of China and Thibet, where a large and thriving trade is done in the wares of both countries. It is also the residence of the King of Chiala, who governs a very considerable tract of country and exercises a strong influence over conterminous states peopled with Thibetans. The first Occidental other than Roman Catholic priests to visit Tachienlu was the late Mr. T. T. Cooper in 1868. Since that date it has been visited by many scores of travellers, and has become fairly well known to the outside world. It is a more than ordinarily interesting place, and though much has been written concerning it the subject is far from being exhausted.

The present town is built in the narrowest of valleys at the head of a gorge, down which the river Lu cascades, falling some 4000 feet before it joins the river Tung, 18 miles distant. A branch of the Lu River bisects the town, being crossed by means of three wooden bridges, and is joined immediately below the north gate by another stream, which flows from the Ta-p'ao shan snows. The town is hemmed in on all sides by steep, treeless mountains whose grassy slopes and bare cliffs lead up to peaks culminating in eternal snow. On the whole, the situation is about the last in the world in which one would expect to find a thriving trade entrepôt. Formerly, Tachienlu occupied a site about half a mile above the present town, but about 100 years ago it was totally destroyed by a landslip, due to a moving glacier. Some day a similar fate will doubtless overtake the existing town.

Notwithstanding its great political and commercial importance Tachienlu is a meanly built and filthy city. It is without a surrounding wall, save for a fragment which runs across near the south gate, and it has no west gate. The narrow, uneven streets are paved with stone in which pure marble largely figures, though this is only evident after some heavy downpour has washed away the usual covering of mud and filth. The houses are low, built of wood resting on foundations of shale rocks. The principal shops are by no means of imposing appearance, and, indeed, the only places noteworthy are two Chinese temples and the palace of the local king. The latter consists of several lofty semi-Chinese buildings of wood with sloping roofs and curved eaves surmounted by gilded pinnacles, the whole structure being situated in a large compound and surrounded by a high stone wall. The residences of the Chinese officials are poor, ramshackle places, and the same is true of the various inns. In the latter most of the business is transacted. Some inns that I visited contained valuable collections of porcelain and bronze-ware, and an extraordinary number of old French clocks. Very few of the clocks were in working order, but many were of large size, and how they all reached this remote place is a mystery to me.

The population of Tachienlu consists of about 700 Thibetan and 400 Chinese families and, with its floating members, is reckoned at 9000 people. In and near the town are eight lamaseries boasting 800 lamas and acolytes. The population is very mixed, consisting of pure Thibetans, pure Chinese, and half-breeds. Very few purely Chinese women are to be found in Tachienlu.

As seen in and around Tachienlu the Thibetans are a picturesque people. Of medium height and lithely but muscularly built, they have an easy carriage and independent mien. The young women are usually sprightly in manner, always cheery, with dark-brown eyes and finely cut features. Both sexes are fond of jewellery ornamented with .turquoise and coral, but they are strangers to soap and water, and personal

TACHIENLU; SITE OF FORMER TOWN IN FOREGROUND

cleanliness is neither appreciated nor practised. Meat, milk, butter, barley-meal, and tea constitute the favourite food of these people ; they are also fond of Chinese wine. Everybody carries on his or her person a private eating-bowl, and the average Thibetan disseminates an odour strongly suggestive of a keg of rancid butter ! The everyday dress of these people is a loose, shapeless garment of dull red or grey woollen serge, sometimes sheep-skins are substituted in part. Top-boots of soft hide with the hair inside usually encase the feet and lower legs of both sexes. The men wear their hair in a queue wound round the head and ornamented with beads and rings of silver, coral, and glass. A large silver ear-ring with a long silver and coral pendant usually decorates the left ear. The women wear their hair parted down the middle and made up into a number of small plaits, which are gathered into a queue, bound at the end by a bright red cord, and wound around the head. Silver and coral are lavishly used in their coiffure and about their persons generally. When in holiday attire these people are more gaily dressed, red-coloured trimmings to their garments being then much in evidence, whilst the wealthy affect silk and fur robes. Ornaments of silver and gold, inset with coral and turquoise, are most profusely worn. The lamas shave their heads and wear a raiment of coarse serge of a dull red or brownish colour. This has no shape, being simply a large piece of cloth thrown over the right shoulder, leaving the left bare. A similar piece of cloth is wound two or three times round the waist and reaches down to the ankle, forming a kind of pleated skirt. They are usually bareheaded and bare-footed, and each lama carries in his hand a rosary and a small praying-cylinder. They swagger through the streets with an insolent mien, and lack the good manners so delightful in the ordinary unsophisticated Thibetan. The lamaseries are usually very richly endowed with land, and most charmingly situated midst groves of Poplar and other trees. Nearly all Thibetan families of affluence maintain a lama on the premises to perform by proxy their religious duties. Many other lamas find employment as temporary chaplains to less wealthy families on occasion of marriage, illness, or death.

Commercially, Tachienlu is a most important centre,

enjoying a monopoly of the trade between this part of China and Thibet. The value of the trade is estimated at about one and three-quarter million taels. The Thibetans bring in musk, wool, deer horns, skins, gold-dust, and various medicines, and take in exchange brick tea and miscellaneous sundries. The trade is largely one of barter, but much less so than that of Sungpan Ting. Sycee and Indian rupees were formerly the only coinage current, but the Chinese during the last few years have been minting in Chengtu a rupee of their own for the special purposes of this trade-centre. Its use has been insisted upon, and, in consequence, the Indian coin has been ousted from the field. Most of the " bigger " trade is in the hands of the lamaseries on the one hand, and Chinese from the province of Shensi on the other. About 30 li to the north-east of Tachienlu gold is found at an altitude of about 11,000 feet, and placer-mining is carried on there. The gold-washing is done in exactly the same way as elsewhere in Western China, but the method of paying the miners is peculiar,—the arrangement being six baskets for the owner of the mine and a seventh for the miners. Silver also occurs at this same place. The Thibetans hold the view that gold and other precious metals grow, and that their death may result if too much is removed at any one time. How far they actually believe in this superstition is a moot point, but at times it serves as an unanswerable argument. Nine years ago a difference of opinion in the matter of assessing the profits arose between the Chief of Chiala, owner of the mine, and the head Chinese official at Tachienlu, who was apparently over-avaricious in the matter. The Chief very quietly advanced the above theory, and closed down the mines for an indefinite period! Gold in great quantity occurs in the state of Litang, west of Tachienlu; much also is mined around Th'ai-ling to the north of this town.

Being on the great highway from Peking via Chengtu to Lhassa, officials are constantly passing through Tachienlu, and the political importance of the town is very great. Although only a city of the second class the head Chinese official has the local rank of Prefect (Chiung-Liang Fu), and holds the post of commissary for the Chinese troops stationed in Thibet. Although Batang, 18 days' journey westwards, is more accurately

the frontier town, Tachienlu is actually the " Gate of Thibet."
The country around and beyond is physically purely Thibetan
in character, and is ruled by native chieftains. Garrisons of
soldiers and a few resident Chinese officials protect the interests
of the Celestial empire and keep a sharp eye on the actions of
the local rulers.

It was stated at the commencement of this chapter that the
King of Chiala resides at Tachienlu, and perhaps a few details
concerning this kingdom and its people may be of interest.
According to the Guide Book of Thibet this State came under
Chinese influence during the Ming Dynasty, about A.D. 1403,
and its Chief was given the rank of a second-class native official,
with control over the tribes west of the river Tung and southwards to Ningyuan Fu. " The Manchu Dynasty, in consideration of the above, made the then Chief a third-class native
official, with power over three trading companies. New chiefs,
chiliarchs, and centurions to the extent of fifty-six were created.
This illustrious Chief now controls six subsidiary chiefs, one
chiliarch, and forty-eight centurions." Since the date of this
appointment the Chinese have increased their grip over these
regions, to the curtailment of the Chief's power and authority.
Nevertheless, the Thibetans of this region acknowledge this
Chief as their supreme ruler, and in domestic affairs his authority
is absolute. His native title is " Chiala Djie-po " (King of
Chiala) ; his Chinese title, " Ming-ching Ssu," which may be
translated " Bright-ruling official." The King and Chinese
Prefect (Fu) are supposed to be colleagues, but in reality the
King is subordinate, and when paying official visits must make
obeisance before the Fu. In what little dealings I had with
them I found both to be courteous and obliging, but suspicious
and jealous of one another.

The present King is a slimly built, intelligent man, about
forty odd years of age. He took considerable interest in our
collecting work around Tachienlu, and with his brother, who is a
hunter of much renown, paid us many unofficial visits. He was
never tired of watching my companion, Mr. Zappey, fixing up
his birds' skins. My own work amongst flowers interested him
but little. As a parting gift Mr. Zappey stuffed and mounted
a Hoopoe for the King, who evinced almost childlike pleasure

on receiving it. In return, he made Mr. Zappey and myself several presents, and urged us to visit his country on a future occasion. We found that these Thibetans possessed keen and accurate knowledge of the birds and animals of their country, which made them enthusiastic hunting companions. During the reign of the former king, his brother, the present Ruler, was banished, and suffered dire hardships during his exile, and often wanted for food. The missionaries stationed in this neighbourhood on more than one occasion assisted him, and I understood from them that he had not forgotten their kindly help. The history of the family is a tragic one. The present King's brother was supposed to have been poisoned, and two sisters died early deaths, the result, it is said, of immoral associations with lamas.

The state of Chiala is of considerable size, comprising practically the whole of the territory lying between the Tung River, Chienchʼiang Valley, and the Yalung River from lat. 28° to 32° N. The five Horba states in a measure also come under the influence of the King of Chiala. From all I can learn this region has the best right to be considered the kingdom of Menia, or " Miniak," of European maps. The whereabouts of " Miniak " has considerably puzzled geographers, but the evidence seems to point to the kingdom of Chiala as representing it in the greater part. North-east of Chiala is the large and prosperous state of Derge, famed for its copper, silver, and swordsmiths. Monsieur Bons dʼAnty, French Consul-General at Chengtu, visited Derge in the autumn of 1910, and on his return gave me a most interesting account of this region. He informed me that Derge is a region of much cultivation, surrounded on three sides by snowclad ranges. The various industries for which the state is famous are not carried on in towns, but by the peasants individually in their homes, and from thence carried to towns for sale. In the valley of the Upper Yalung, abutting on the north-west frontier of Chiala and the south-east frontier of Derge, is a wedge of country known as Chantui, peopled by a race of Ishmaels, whose hands are ever turned in conflict against their neighbours. A similar people occupy a wedge of country in the Drechu valley north of Batang, where they are known as

MANI STONES

the Sanai tribe. Monsieur Bons d'Anty considers that these people are of Shan origin, and remnants of an aboriginal population of this region. This authority has spent many years in studying the ethnological problems of this borderland, and is most competent to express an opinion. It is well known that the Shans formerly ruled in western Yunnan, and there is no reason why they should not, in the distant past, have ascended the valleys of the Yalung and Drechu and established themselves there. But whatever the origin of these people of Chantui and Sanai, they are dreaded by their neighbours, who regard them all as robbers and murderers (Ja-ba) quite beyond the pale.

The religion of the people of Chiala is Lamaism, both the orthodox " yellow " and unorthodox " red " sects being represented, but the former are the more numerous and powerful. Some one has described Lamaism as " mechanical," a most descriptive term, since the religion consists in the main of turning praying-wheels by hand, water, or wind, counting beads, and the continual muttering or chanting of the mystic hymn, " Om mani padmi hum." Lamaism draws its inspiration from Lhassa, where all the priests repair for study, the head of the sect being the Dalai Lama. Aided and abetted by Chinese authority, the King of Chiala has never submitted to the Dalai Lama in temporal affairs ; he has maintained his freedom and right to govern his own people untrammelled by Lhassa interference, in spite of the dire threats and treachery on the part of lamaseries within his jurisdiction. In 1903 the Dalai Lama issued an ultimatum to the King of Chiala threatening to take from him and the Chinese by conquest all the territory west of the Tung Valley. The British Expedition prevented the carrying out of this threat. The Dalai Lama undoubtedly had designs of territorial expansion at the expense of China's vassal states. The Chinese knew this, and it was fortunate for them that Great Britain stepped in and broke the power of Lhassa De. I was in Tachienlu during 1903 and 1904, and from what I saw and heard there it was plain that the British were unwittingly pulling China's " chestnuts from the fire." The Chinese were not slow to perceive the advantageous position they were in after the power of the Dalai

Lama was dissipated. Almost immediately a "Wardenship of the Thibetan Marches" was established, and a war of conquest engaged upon against certain wealthy lamaseries in Litang and other states, who owned direct allegiance to Lhassa, and heretofore had boasted their independence of China. This war was relentlessly and victoriously pursued under the leadership of Chao Êrh-fêng, and resulted in the extension of Chinese authority over a very considerable tract of country. Indirectly the King of Chiala's position has been very much weakened as the outcome of these conquests.

The state of Chiala is made up of mountain, dale, and plateau, being essentially a highland country affording good pasturage for yak, sheep, and horses. A chain of snowclad peaks traverses it near its eastern boundaries. It is a region where altitude regulates the mode of life, the wealth, and marriage customs of its people. The inhabitants are less nomadic than the people to the north and west, but, in common with all other Thibetans, their wealth is represented by herds of yak, horses, and cattle, and flocks of sheep and goats. They are great hunters of Musk-deer, Wapiti, Bear, and other animals, the commercial products of which they trade to the Chinese. The same is true of the medicinal roots and herbs, which grow abundantly in these uplands. Where altitudes admit, agriculture is practised, but is supplementary to grazing and relatively unimportant. Wheat, barley, oats, buckwheat, peas, and Irish potato are the chief crops. During the winter months these Thibetans live in well-built houses situated in the valleys, and in the spring they migrate to the uplands. The nomads do not move about aimlessly, but have clearly defined regions and are subject to responsible head-men. Where agriculture is carried on the womenfolk mostly remain to look after the crops and to do other work pertaining to the farmstead.

Wealth and convenience decide which form the matrimonial alliance shall assume among these people, and polygamy, monogamy, and polyandry obtain. Above 12,000 feet altitude polyandry is the rule, and in many places women so united wear distinguishing and honorary badges. Such women are usually the business and ruling heads of their establishments. This custom of polyandry is characteristic of Thibet, and the following note on the subject, written by a friend who has spent many years of his life among these people, is worthy of much thoughtful study :—

"So many able men have written about polyandry that what follows will be without interest to those who have studied the system ; but to the great mass who are comparatively unacquainted with Thibet and her customs these notes may be of some value. The writer has spent several years among Thibetans and cognate tribes, and has lived for months alone on the wild steppes as well as in the more civilized and well-cultivated valleys.

"The term ' polyandry ' is here applied (a) to women living permanently, and cohabiting legally, with more than one man ; (b) to those who have been, or are, married temporarily to more than one man or companies of men.

"The former, true polyandry, is confined to the pastoral nomads of the grassy plateaux ; the latter, quasi-polyandry, is rampant in all the commercial and political centres on the border and throughout Thibet. In both cases a low conception of the relation of the sexes has made it possible ; and climate and political conditions have made it desirable.

"The past hints, and the present proves, that indifference to female virtue connotes the people known as Thibetans and tribes of common origin, and I understand it to be the indirect cause of polyandry. From time immemorial the Thibetan has been taught that the female is a kind of Pandora's box, in which are all the evils that have cursed mankind. All down the ages woman seems to have been the slave of man : dangerous because of latent evil, but also valuable on account of her ability to render him service. In the old barbaric days, when prowess was the prime virtue and a thoroughgoing communism the rule, woman was only a tribal asset, like the animals she tended. Then came religion, a deification of all that rude minds could not explain. It was probably the mysterious Bönpa of to-day which lingers in the lonely valleys where nations meet, and which could have been no friend of virtue if the accounts of orgies in its temples before indecent idols are true, and the unseemly dress of young women and barren wives either demanded or sanctioned by it. Explain it as we may, the fact remains that Thibetan women are to-day, as they seem to have been in the time of Marco Polo, the most immodest of their sex, and the Thibetan men strangely indifferent about matters which other races demand as essentials.

" All outside work is done by the women, who represent the coarser element of Thibetan society, and their language is often filthy in the extreme. The domestic arrangements make no provision for privacy. Men and women must eat, live, and sleep perforce in the same apartment, and there is no effort on the part of the male to shield the female from conditions which are inimical to virtue.

" The morality of the Thibetans has made such a system possible. This will not be denied by any one who knows them even slightly ; but it will sound strange to many when I say that the climatic and political conditions are such that the reformer is puzzled to think of anything to offer as a substitute ! To the untutored Thibetan mind it must seem absolutely necessary. Undoubtedly the high altitudes are unfavourable to women. The Thibetan views woman very much as he does an animal, i.e. she can do so much work. Living and working at 12,000 feet altitude and upwards requires the strongest material. Woman very imperfectly fulfils these requirements, and maternity and nursing, apart from unfitting her for work, would be well-nigh useless, since infant mortality would be abnormally high. On the relatively thinly populated plateaux the conditions obtaining are emphatically against woman being wanted in numbers. Here robbing and escaping from robbers is the normal condition. It will be evident at once that family duties are not only inconvenient, but interfere with the woman's efficiency personally, and at the same time misdirect the energies of the male portion of the community.

" The nomad is a herdsman, continually moving to and fro with his flocks and belongings. The woman, and the centre she forms, would impair the necessary freedom of movement ; it would also follow that she and her belongings would often be unprotected for long periods. Polyandry, by not encouraging permanent settlements and at the same time being the best

security against marauding bands, must seem eminently
rational to the nomad.

" Polyandry also entails the family property. This is very
important, as division of the flocks or grazing-grounds would
soon ruin every one. Whatever the ideal system for these
Thibetans may be, the one which provides one wife, one family,
and one flock for all the male members of the family is the most
convenient. Anything else would be suicidal. Both polygamy
and monogamy presuppose racial increase and the formation
of new and independent centres, but polyandry promises the
great desideratum of the Thibetan—an almost stationary
community and an intact patrimony.

" In a land of polyandry, priestly celibacy, and nondescript
roving, the number of unmarried women must be large. This
class, with the Chinese, Lamas, and Thibetan merchants, is
responsible for the quasi-polyandry of the plain, which only
differs from prostitution inasmuch as it has the sanction of the
country and carries with it no odium. The priest is a celibate,
as a rule, by profession, but an inveterate roué in practice.
Quite a large number of women are required wherever lamaseries
exist. In Lhassa, where thousands of students from all parts
of the country study for years, the number of women married
temporarily, openly, or in secret, to individuals or small communities is very great. The wandering Thibetan merchants
form another class who demand a supply of temporary wives for
longer or shorter periods. These may often be men who have
formed polyandrous unions in the mountains, but the exigencies
of circumstances demand their presence on the plain. In other
words, there is no reason why a man may not be a polyandrian
legally, and in practice a polygamist.

" But the most interesting phase of this system arises from
peculiarities of Chinese domination. Chinese soldiers, officials,
and merchants residing temporarily in Thibet form a very
large body. These victims of circumstances leave their wives
in far-away China. There is a legend that the Lamas have put
an embargo on the dainty Chinese woman : but, more probably, her lord and owner has neither the mind nor the money
to introduce her to the dangers and hardships of a Thibetan
journey. But he rarely, if ever, pines for the wife of his

youth. Polyandry and polygamy meet, and temporary marriages, for one month to three years, are the rule. The highest official and the meanest soldier take advantage of the system. With the former it is temporary monogamy or polygamy, but with the latter, owing to pecuniary limitations, one woman often becomes, *pro tempore*, the wife of a small community of soldiers. These wives or their children, for obvious reasons, are seldom, if ever, brought out from Thibet ; the former make new alliances and the children are claimed by the Lamas.

" The question of Thibetan morality is a very complex one, and it is almost impossible to disentangle the cause from the effect. True polyandry is owing, indirectly, to a low moral perception ; but it might be correct to blame it, in a measure, for the more degenerate quasi-polyandry. Whatever we may think of the former, from the standpoint of absolute morality, it is relatively a moral system and solves many problems. To change it without changing the conditions would be tantamount to driving the brave nomad women into the towns to become the temporary wives of Chinese rabble, priestly roués, and peripatetic Thibetans. Perhaps my hinting that polyandry as a system is in many ways well suited to the plateaux will evoke much unfavourable comment, but there are good men, Roman Catholic and Protestant, priest and layman, who have noticed the same difficulty.

" The effect of the system on the women is another question about which we cannot afford to be dogmatic. When young the Thibetan women are often very pretty, but they age quickly and become as weirdly ugly as the mediaeval witches. To say that polyandry is alone responsible for this change would be sentiment unsupported by facts ; but undoubtedly this system, combined with hard work, loathsome uncleanliness, and often grotesque head-dress tends to give a great many women an inhumanly vile expression.

" The families on the plateaux are very small and many women are barren. This is a blessing in disguise, owing to the impossibility of the nomad country supporting more than a very limited population, and the small amount of arable land capable of relieving the congested centres. Polyandry is both

THE LAMASERIES JUST OUTSIDE TACHIENLU

directly and indirectly the cause of this limitation of offspring. A glance at the system will show how these uncultured Malthusians obtain their end : Three men, for instance, centre their affections on one woman, who in her lifetime rears two or three children. As monogamists each of these men would have had his own wife and probably a total of fifteen children. But another factor has to be taken into consideration : polyandry not only limits a woman's natural fecundity, but in a great number of cases is the direct cause of barrenness.

" About the domestic arrangements I cannot speak authoritatively, but I have never heard internal discord used as an argument against polyandry. It must often happen that one or two husbands are away tending flocks, worshipping at holy mountains, or robbing travellers. But this is an accident ; the domestic equilibrium is rarely disturbed by petty jealousies. The defloration of the bride or brides— for there is no reason why two or more sisters should not come into the community—is the right of the elder brother, and the first child is, by courtesy, assigned to him ; but the child or children of the union are, in reality, a joint possession. The girls in the community either follow their mother's example, or go into the towns and become the temporary wives of Chinese, Lamas, or wandering merchants. In the former case a dowry is given to the parents, but in the latter the ' fair one ' makes the most of her time and the simplicity of her husband or husbands.

" Polyandry in one form or other is probably practised whenever Thibetan communities are found. Its existence may be denied emphatically, but closer investigation will only prove the wide distribution of the ' Münchausen ' family. However, an exception may be allowed in the deep, populous valleys of Eastern Thibet. Here individualism is the rule, and new centres are formed and thrive without the shadow of a grim Frankenstein disturbing them. So completely has the old dread of offspring been effaced that marriage is always preceded by a tentative period, and maternity alone establishes a girl's right to be admitted into her husband's family. Here the quondam upholder of polyandry, realizing that the fruitful earth and the fruitful woman bring wealth and strength respectively, becomes a confirmed polygamist. To the student of ethnology this metamorphosis suggests the permanency of the valley Thibetan and the gradual absorption or total extinction of his mountain brother."

CHAPTER XVII

SACRED OMEI SHAN

Its Temples and its Flora

THE lofty and sacred eminence known as Mount Omei, or Omei shan, is situated about long. 103° 41′ E., lat. 29° 32′ N., one day's journey from the city of Kiating. A gigantic upthrust of hard limestone, it rises sheer from the plain (alt. 1300 feet) to a height of nearly 11,000 above sea-level. From the city of Kiating a fine view of this remarkable mountain is obtainable during clear weather, the mirage of the plain seemingly lending it additional height. Viewed from a distance it has been aptly likened to a couchant lion decapitated close to the shoulders, the fore-feet remaining in position. The down-cleft surface forms a fearful, well-nigh vertical precipice, considerably over a mile in height ! It is one of the five ultra-sacred mountains of China, but the origin of its holy character is lost in antiquity. We are told that in a monastery here the patriarch P'u (an historical personage) served Buddha during the Western Ts'in Dynasty (A.D. 265–317). P'u-hsien Pu'ssa (Samantabhadra Bodhisattva), Mount Omei's patron saint, descended upon the mountain from the back of a gigantic elephant possessed of six tusks. In one of the temples (Wan-nien-ssu) there is a life-sized elephant cast in bronze of splendid workmanship which commemorates this manifestation. Upwards of seventy Buddhist temples or monasteries (either word is applicable, since the buildings are really a combination of both) are to be found on this mountain. On the main road to the summit there is a temple every 5 li, and they become even more numerous as the ascent finally nears the end. These temples are controlled by abbots and contain upwards of 2000 priests and acolytes. The whole of the mountain is, or rather was, church property, much of the land on the lower slopes suitable for cultivation having from time to time been sold away from the church. Voluntary subscriptions are now the chief sources of revenue of the religious houses, though many of the temples have money as well as land endowments.

Many thousands of pilgrims, coming from all parts of the Chinese Empire, visit this mountain annually. At the time of my visit there were several pilgrims who had walked all the way from Shanghai, some 2000 miles distant, for the express purpose of doing homage before the shrines of Mount Omei. Thibetans and even Nepalese make pilgrimages here. The images and sacred objects are numberless, many of them being of pure bronze or copper. Three mummified holy men, lacquered, gilded, and deified, the elephant above mentioned, and a tooth of Buddha are among the more interesting objects. The tooth is about a foot long and weighs 18 English lb., and is in all probability a fossil-elephant's molar. On the extreme summit of the mountain, the Golden Summit, as it is called, are the ruins of an ancient temple which was built of pure bronze. It is said to have been erected by the Emperor Wan-li (A.D. 1573–1620), and was destroyed by lightning in 1819. Since this catastrophe nine or ten abbots have come and gone, but none has been able to collect enough money to rebuild it. The mass of metal at present heaped around, consisting of pillars, beams, panels, and tiles, is all of bronze. The panels are particularly fine pieces of work. I measured one panel which had dimensions as follows : 76 inches high, 20 inches wide, $1\frac{1}{2}$ inches thick ; some of the panels are slightly smaller than this. All are ornamented with figures representing seated Buddhas, flowers; and scroll-work, and on the reverse with hexagonal arabesques. Many of the panels have been incorporated in one of the two small temples which now stand on the crest of the precipice. Wan-li's tablet, which was contained in the ancient bronze temple, is to-day accommodated in an outhouse along with fuel. The crown-piece is detached and lies outside. This tablet is of bronze, but is hollow. With crown-piece and pedestal it measures 90 inches high, 32 inches wide, and 7 inches thick. Another grand relic

THE CHINESE BANYAN (FICUS INFECTORIA) 70 FT. TALL, GIRTH 47 FT.

left to the tender mercies of the elements is a huge bell which
stands 54 inches high and is 120 inches round the middle.
On the edge of the cliff are two bronze pagodas, each about
12 feet high, and the remains of a third, which formed
part of the ancient temple. It is a saddening sight to gaze
around on these most interesting relics so ignominiously
neglected.

From the summit of Mount Omei, when the sky is clear and
clouds of mist float in the abyss below, a natural phenomenon
similar to that of the Spectre of the Brocken is observable. I
have never seen it myself, since rain fell almost continuously
during the week I spent on the summit, but it has been described
as a " golden ball surrounded by a rainbow floating on the
surface of the mists." This phenomenon is known as the
" Fo-kuang " (=" Glory of Buddha "). Devotees assert that it
is an emanation from the aureole of Buddha and an outward
and visible sign of the holiness of Mount Omei. The edge of the
precipice is guarded by chains and wooden posts, but pilgrims
in a state of religious fervour have been known to throw
themselves over on beholding the Fo-kuang. From this cause
the point is called the " Suicide's Cliff." It is the highest and
most vertical part of the precipice, which extends in a nearly
southerly direction for a couple of miles.

The first foreigner to ascend this famous mountain was
the late E. Colborne Baber, who visited it in July 1877, and
whose incomparable and accurate account of this region has
never been equalled.[1] Unfortunately Baber paid little or
no attention to the flora, nor did the equally distinguished
traveller and writer Hosie,[2] who ascended Omei shan in 1884.
It was not until 1887 that any plants were collected on this
mountain. In that year it was visited by a Rhenish missionary,
who was also an industrious botanical collector—the late Dr.
Ernst Faber. During a fortnight's stay this enthusiast made
a most interesting collection; which was found on critical
examination to contain no fewer than seventy novelties. In
1890 an English naturalist, Mr. A. E. Pratt, visited the mountain and collected a few plants. Since Baber's visit many

[1] *Royal Geographical Society, Supplementary Papers,* vol. i.
[2] Sir Alexander Hosie, K.C.M.G., H.B.M.'s Consular Service in China.

hundreds of foreigners have ascended Mount Omei, but with the exception of those of Faber and Pratt, there is no record of any one having collected plants during their visits. For this reason alone I hope this chapter will find justification. The mountain and its temples have been well described by Baber and others, and I have no desire to attempt to repeat descriptions which have been made by abler pens than mine. With this prelude I append the following record of my visit :—

It was on the morning of 13th October 1903 that I set out from the city of Kiating intent on investigating the flora of this famous mountain. Traversing the highly cultivated plain, which is intersected here and there by low hills, charmingly wooded, the little town of Omei Hsien (alt. 1270 feet) was reached at the close of the day. The next morning, after journeying 10 li across the plain along a road shaded with trees of Alder and Nanmu, we reached the village of Liang-ho-kou, situated at the foot of the sacred mountain. Here the road bifurcates and both paths lead by different routes to the summit. They are paved with blocks of stone throughout, an undertaking that must have entailed a vast expenditure in labour and money, but it would be impossible to traverse certain of the steeper parts unless paving existed. I ascended by one of the routes and returned by the other, so that I saw as much as was possible of the mountain and its rich flora.

Between Omei Hsien and Liang-ho-kou are a number of truly magnificent Banyan trees (*Ficus infectoria*), known locally as Huang-kou-shu. These trees shelter some old temples and are of enormous size. I measured one, which appeared to be the largest specimen ; it was about 80 feet tall, and had a girth of 48 feet at 5 feet from the ground. We also passed some fine trees of Oak (*Quercus serrata*) and Sweet Gum (*Liquidambar formosana*). The sides of the rice fields are studded with thousands of pollarded trees of the Chinese Ash (*Fraxinus chinensis*) on which an insect deposits a valuable white wax. The ditches were gay with the spikes of cream-coloured, fragrant flowers of a species of Hedychium, the golden-flowered *Senecio clivorum*, flowers of many kinds of Impatiens, and other moisture-loving herbs.

On leaving Liang-ho-kou the ascent began, and journeying slowly three days' hard climbing brought us to the " Golden Summit."

For the purpose of grouping the flora it is convenient to divide the mountain into two regions—(1) from the base to 6000 feet, and (2) 6000 feet to the summit (10,800 feet). Thus divided the flora falls into two well-defined altitudinal zones. The lower zone is made up of such plants as enjoy a warm-temperate climate. Evergreen trees and shrubs predominate, and in the shady glens and ravines Selaginellas and Ferns luxuriate. Of these latter I, in one day, collected over sixty species ! The upper zone consists entirely of plants requiring a cool-temperate climate. With the exception of Rhododendron and Silver Fir it is composed almost entirely of deciduous trees and shrubs and herbaceous plants. The belt between 4500 feet and 5500 feet may be termed the Hinterland. Here the struggle for supremacy is most keen and the fusion of the zones most marked. At 6000 feet the boundary line is unusually well defined.

Cultivation extends up to 4000 feet, maize and pulse being the principal crops, with rice relegated to the valleys and bottom-lands. Plantations of Ash trees for the culture of insect-wax extend up to 2600 feet. The foot-hills around the base of the mountain are covered with Pine (*Pinus Massoniana*), Cypress (*Cupressus funebris*), and Oak (*Quercus serrata*). The sides of the streams which meander among these hills are clothed with Alder (*Alnus cremastogyne*), *Pterocarya stenoptera*, and the curious *Camptotheca acuminata*. Around the temples and farmsteads Nanmu and tall-growing Bamboos abound ; on the more exposed hillsides the climbing fern *Gleichenia linearis* forms impenetrable thickets, and *Onychium japonicum*, *Melastoma candida*, *Mussœndra pubescens*, are common road-side plants. At 3000 feet all these plants drop out and give place to others. *Cunninghamia lanceolata*, which occurs sparingly in the valleys, gradually increases in number, and between 2500 and 4500 feet large areas are covered solely with this invaluable Conifer. Apart from the Cunninghamia, the family of *Laurineœ* forms, between 2000 and 5000 feet, fully 75 per cent. of the arborescent vegetation. This " Laurel zone," as it may be termed, is composed chiefly of evergreen trees and shrubs, the genera Machilus, Lindera, and Litsea being exceptionally rich in species. Within this zone also occur the following interesting monotypic trees : Tapiscia, Carrieria, Itoa, Emmenopterys, and Idesia. The evergreen *Viburnum coriaceum*, with blue-black fruits, and five species of evergreen Barberries are also met with found here.

In ascending any high mountain, more especially in these latitudes, it is most instructive and interesting to note the aggressiveness of the temperate flora. Mount Omei offers special facilities for studying this phenomenon. Everything around us looks so smiling that all nature seems to be at peace. In these days, however, every one is alive to the fact that a stern and relentless war of conquest is being continually waged on all sides, and that every inch of ground is contested. It is well that plants cannot speak, or the exultations of the victors and the groans of the vanquished would be too much for humanity to bear !   But to note the struggle : The large-leaved Cornel (*Cornus macrophylla*) manages to extend its area nearly to the base of the mountain, being closely attended by several species of Maple, among which *Acer Davidii*, with white-striped bark, is particularly prominent. A Black Birch (*Betula luminifera*), several species of Viburnum, Pyrus, Malus, Rubus, and Prunus are also well to the fore ; but it is in the Hinterland (4500 to 5500 feet) that the main battle between the zones is fought. This narrow belt is extraordinarily rich in woody plants. Of those peculiar to it I may mention *Pterostyrax hispidus, Pterocarya Delavayi, Euptelea pleiosperma, Decaisnea Fargesii*, Horse Chestnut (*Æsculus Wilsonii*), and the monotypic genera Tetracentron, Emmenopterys, and Davidia. At least five species of Maple occur with many fine specimens of each. Several species of Evonymus, Holbœllia, Actinidia, and Holly (*Ilex*) are also common. The bulk of the *Laurineæ* have given up the struggle, their place being taken by Evergreen Oak and Castanopsis. In this belt monkeys are common, and are fond of the blue pod-like fruit of the Decaisnea, the shining black, flattened seeds of which, however, I noticed they cannot digest.

On clearing a dense thicket and emerging on to a narrow

HYDRANGEA XANTHONEURA VAR. WILSONII, 15 FT. TALL

ridge, 6100 feet above sea-level, a magnificent view presented itself. Above towered gigantic limestone cliffs nearly a mile high ; below spread valleys and plains filled with a dense, fleecy cumulus, through which the peaks of mountains peered like rocky islands from the ocean's bed ; to the westward the mighty snowclad ranges of the Thibetan border, 80 miles distant as the crow flies, presented a magnificent panorama stretching northward and southward as far as the eye could range. The contrast between the floral zones was equally startling and impressive. Below, until lost in the clouds, was a mass of rich, sombre, green vegetation ; above were autumnal tints of every hue, from pale yellow to the richest shades of crimson, relieved by clumps of dark green Silver Fir. The whole scene was bathed in sunlight, a gentle zephyr stirred the air, and gorgeous butterflies flitted here and there seemingly unconscious of winter's near approach. The stillness and quiet was most solemn, and was broken only by the warbling of an occasional songster in some adjacent tree or bush. It was indeed a never-to-be-forgotten scene !

At 6200 feet the Cunninghamia gives up the fight, having struggled nobly until reduced to the dimensions of an insignificant shrub. A Silver Fir (*Abies Delavayi*) next assumes the sway, and right royally does it deserve the sceptre, for no more handsome Conifer exists in all the Far East ; its large, erect, symmetrical cones are violet-black in colour and are usually borne in greatest profusion on the topmost branches. The temples on the higher parts of the mountain are constructed almost entirely of the timber of this tree. It is first met with on Mount Omei, at 6000 feet, at which altitude it is of no great size and unattractive in appearance ; at 6500 feet it is a handsome tree. It is, however, between 8500 and 10,000 feet that this Silver Fir reaches its maximum size. In this belt hundreds of trees 80 to 100 feet tall, with a girth of 10 to 12 feet, are to be found. Hemlock Spruce (*Tsuga yunnanensis*) occurs sparingly, but always in the form of large and shapely trees. An occasional Yew tree (*Taxus cuspidata*, var. *chinensis*) and, on the summit, dwarf Juniper (*J. squamata*) complete the list of Conifers growing on the higher parts of this mountain. The unspeakably magnificent autumnal tints already

referred to are principally due to numerous species of
Viburnum, Vitis, Malus, Sorbus, Pyrus, and Acer, together
with *Enkianthus deflexus*, which surpasses all in the richness
of its autumn tints of orange and crimson.

At 6200 feet the ascent becomes increasingly difficult,
and having surmounted a formidable flight of steps, 800 feet
high, we were glad to rest at the temple of Hsih-hsiang-chüh.
All the temples on Mount Omei occupy lovely and romantic
situations, but none more so than this, which has one side
flush with the edge of a precipice, and the others sheltered by a
grove of Silver Fir.   The hospitable priests regaled us with tea
and sweetmeats and entertained us with much that was curious
and amusing.   They claimed that it was at this particular
place that P'u-hsien Pu'ssa alighted from his elephant to allow
the footsore animal to bathe in a near-by pool; the spot
to-day is marked by a cistern.

Immediately on leaving this temple two steep flights of steps,
followed by a slight descent, led us to a small wooded plateau
which shelves away from a vertical precipice.   Hereabouts
Sorbus munda, with white fruits, was a most conspicuous
shrub.   A climbing Hydrangea (*H. anomala*) reaches to the top
of the tallest trees.   Several other species of Hydrangea grow
epiphytically on the larger trees and so also do two or three
species of Sorbus.   Rhododendrons are fairly abundant,
more especially near the edge of the precipice.   The first few
Rhododendron bushes were noted growing at 4800 feet, and
altogether I gathered thirteen species on this mountain.
But as compared with the region to the westward Mount Omei
is poor in Rhododendrons.   The same is true of Primulas,
of which four species only were met with.

At 9000 feet the most difficult stairway of all occurs, and
I was fairly exhausted when the top of it was reached at
10,100 feet.   Winter had laid his stern hand heavily here,
and most of the woody plants were leafless.   At 10,000 feet
Bamboo-scrub puts in an appearance and increases as the
summit is neared until finally it crowds out nearly everything
else and forms an impenetrable jungle about 4 to 6 feet high.

From the top of the last stairway an easy pathway of
planking leads to the summit, which we reached just as the

sun was setting behind the snowclad ranges of the Thibetan border.

A perfect night succeeded the day, and our hopes were high for the morrow. Alas ! a thick fog and a drizzle of rain was what we awoke to find. A terrible precipice in front and a more or less shelving away behind was all we could make out of the lay of the land. To find out what the summit is really like, a long walk was undertaken, but resulted in little beyond a thorough drenching. The mountain-top is somewhat uneven, sloping away from the cliffs by a fairly easy gradient. It is everywhere covered with a dense scrub, composed mainly of dwarf Bamboo, with bushes of Willow, Birch, Sorbus, Barberry, Rhododendron, Spiræa, and *Rosa omeiensis* interspersed. Near the watercourses these shrubs are more particularly abundant. Trailing over the scrub *Clematis montana*, var. *Wilsonii*, is very common. At least five species of Rhododendron grow on the summit, but, judging from the paucity of fruits, they flower but sparingly. In places sheltered from the winds fine groves of Silver Fir remain, but in the more fully exposed sites these trees are very stunted and weather-beaten. The dwarf Juniper, with twisted, gnarled stems, is also plentiful in rocky places.

Around the temples small patches of cabbage, turnips, and Irish potato are cultivated, and several favourite medicines are grown in quantity, such as Rhubarb, " Huang-lien " (*Coptis chinensis*), " Tang-shên," and " Tang-kuei."

Here and there on the mountain we passed hucksters' stalls, on which various local products were exposed for sale. These consist chiefly of medicines, porcupine quills, crystals of felspar, sweet tea, and pilgrim staves. The latter, made from the wood of an Alder (*Alnus cremastogyne*), are carved in representation of fantastic dragons and Buddhas. The sweet tea is a peculiarity of Mount Omei, being prepared from the leaves of *Viburnum theiferum.*

CHAPTER XVIII

THROUGH THE LAOLIN (WILDERNESS)

Narrative of a Journey from Kiating to Malie, via Wa-wu Shan

LEAVING the city of Kiating on 4th September 1908, we followed the main road to Yachou Fu and stayed for the night at Kiakiang Hsien, a small city, altitude 1200 feet, 70 li from our starting-place. It had rained heavily in the early morning, but cleared just before we set out, and was cool and fine, although dull the whole day. The road is broad, mostly well paved, and leads through a rich and highly cultivated region. Around Kiating the rice had been harvested, much of the land reploughed, and another crop, chiefly buckwheat and turnips, planted. A few miles beyond this city, however, the rice crop was not so forward, and though a portion was being reaped the bulk would not be ripe for some weeks.

Around the margins of rice fields trees for the culture of insect white-wax are abundantly planted. Pollarded Ash (*Fraxinus chinensis*) were chiefly noticeable, but in places trees of Privet (*Ligustrum lucidum*) are used for this industry. Much of the wax had been collected, but in one place we were fortunate enough to witness the process and obtain photographs. (In Vol. II., Chapter X., this interesting industry is fully described.) Sericulture was very much in evidence, and all the alluvial flats are planted with Mulberry trees, but trees of Cudrania are not common. In this region in particular the silkworms are fed on the leaves of both these trees ; the people claim that this mixed diet results in a stronger kind of silk.

The Szechuan Banyan (*Ficus infectoria*) is the most striking tree hereabouts ; its widespreading umbrageous head usually shelters some wayside shrine. Venders of cakes, pea-nuts, and fruit are also to be found occupying some temporary stall under these beautiful trees. The road skirts the sides of low hills of red sandstone for considerable distances, and is mainly parallel with, and in full view of, the Ya River. The hills are clad with common Pine (*Pinus Massoniana*), Cypress (*Cupressus funebris*), a jungle-growth of low shrubs, and the scandent *Gleichenia linearis*. Small trees of Oak and Sweet Chestnut and larger ones of Alder are also common. Groves of tall-growing Bamboos, of course, are everywhere abundant. In the sandstone cliffs are very many square-mouthed Mantzu caves ; the scenery is distinctly pretty and pleasing.

We left Kiakiang at 6.30 a.m. the following day, and quickly reached a ferry, where we crossed over the Ya River, a broad, stony, shallow stream. Quite near this place are two really fine and very large old temples known as Ping-ling-ssu and Kuei-ling-ssu. The first named, in particular, contains some very fine idols ; both, however, have a very deserted and neglected appearance, and give the impression of " glories departed." The sandstone cliffs at the ferry are highly sculptured, but are rapidly weathering away, much of the work being undecipherable and hidden by vegetation.

The li proved very long, and we did not reach Che-ho-kai until 7 p.m., going steadily the whole day. The distance is 80 li, and three ferries, which hinder considerably, have to be crossed. Near the city of Hungya Hsien, which we sighted in the late afternoon, large plantations of Ash trees for the culture of insect white-wax abound. Rice is everywhere the great crop ; the yield was heavier than usual, and the people were busy reaping and threshing it. Fine Banyan trees are plentiful, Alder is abundant, and handsome Nanmu trees are not infrequent around temples and houses. We also noted a small tree of the Hog-plum (*Spondias axillaris*) bearing quantities of its oblong, yellow, edible fruits. The vegetation generally is similar to that around Kiating, but the Chinese Fir (*Cunninghamia lanceolata*) is more common and Pine and Cypress less so.

Che-ho-kai, alt. 1400 feet, is a large and important market village, situated on the right bank of the Ya River. The inn is very fair. I occupied a large room overlooking the river, but, as I discovered later, with a piggery and latrine below.

The next day we began our real journey. Instead of following the main route to Yachou Fu by crossing over the river, we ascended the right bank for a couple of li beyond Che-ho-kai, and then crossed a considerable affluent of the main stream. Rafts of good-sized poles of Chinese Fir descend this tributary from Liu ch'ang, a market village, and ordinary bamboo-rafts ascend to this place. After climbing to the tops of some low hills the road zigzags around considerably through fields of rice and wooded knolls, and affords an unusually fine view of the Ya Valley. Passing the tiny market village of Tung-to ch'ang we reached Kuang-yin pu (or ch'ang) at 10.45 a.m., having covered 30 li.

From Kuang-yin pu we engaged in a steep ascent over a well-paved if narrow road, and after four hours' climbing reached the summit of the Fung-hoa-tsze, alt. 4100 feet. This ridge is of red sandstone throughout, and is well timbered with small trees of the Chinese Fir. This conifer abounds on the slopes flanking the roadway to the top of the pass and forms pure woods. Though the timber is of no great size, the area covered with this tree compares most favourably with any other I have seen. Where timber is scarce the jungle growth is very thick, warm-temperate in character, and of little interest.

Descending, at first steadily, through knolls covered with Chinese Fir and the densest fern jungle composed of *Gleichenia linearis* I have ever seen, we soon reached an area under maize. From this point a steep descent led to a cultivated flat, then, after winding through rice fields with tiny wooded hillocks on all sides, we crossed a neck and entered the hamlet of Liang-ch'a Ho, alt. 2350 feet, and 65 long li from our starting-point. We found very decent accommodation, all things considered, but mosquitoes were most unpleasantly numerous and hungry.

It rained very heavily during the early morning of the next day, so we delayed our starting until eleven o'clock. We found all the streams in flood, and to cross one larger than the ordinary we had to engage local assistance. After a

TEMPLE ON THE CITY WALL, KIATING FU

rather steep ascent of 500 feet from Liang-ch'a Ho we crossed
a narrow ridge and descended to the market village of N'gan
ch'ang. This is a poor place, partly in ruins, situated on
the right bank of the stream which unites with the Ya Ho,
2 li above Che-ho-kai. On leaving N'gan ch'ang we ascended
the right bank of the stream to Pao-tien-pa, alt. 2600 feet.
This scattered hamlet possesses no inn, but we found quarters
in a schoolhouse devoted to the "New Learning" (*i.e.*
Western Knowledge). A scholar from this place had recently
gone to Japan to increase his store of knowledge, and the
dominie was very proud of this success. This hamlet boasts
a ruined pavilion, a temple, and a stone gateway, evident
signs of former prosperity.

During the short journey of 25 li the road led through
fields of rice, bounded by wooded knolls and sandstone
bluffs. The flora was of little interest ; *Idesia polycarpa* and
Kalopanax ricinifolium are fairly common in places, but the
trees are of small size. Alongside the ditches and roadway
the handsome *Lycoris aurea* abounds, and the golden-yellow
flowers with recurved, wrinkled, perianth-segments made
a gay display. Its red-flowered counterpart, *L. radiata*, also
occurs, but is much less frequent. The local name for this
plant is "Lao-wa-suan," which signifies "Crow's foot Onion,"
a very apt term in so far as the shape of the flower is concerned.

The following day was fine but hot, and more or less
cloudy. With only 35 li to cover, we journeyed slowly after
making an early morning start. A moderately steep ascent
of 15 li brought us to the summit of the Tsao shan, alt.
4100 feet. This ridge is covered with an uninteresting jungle
of coarse grass and scrub, with odd trees of Chinese Fir, but
in the ascent I gathered specimens of a fine new species of
Castanopsis.

From the summit of Tsao shan we obtained our first view
of the Wa-wu shan, an extraordinary-looking massive mountain,
singularly like Wa shan in contour, resembling a huge ark
floating above clouds of mist. Following an easy path which
led through fine woods of Evergreen Oak, Nanmu, and
Castanopsis we descended to Ma-chiao-kou, where there is an

iron suspension bridge over a wide torrent. This hamlet consists of one large house and a mill, where a specially good and tough bamboo paper is made, which is used at Yachou Fu for wrapping up brick tea. The bamboo is obtained from the surrounding mountains, and is a species with dull green culms about the thickness of a man's thumb, growing 12 to 15 feet tall. On crossing over the bridge, I photographed a fine specimen of *Alniphyllum Fortunei*, one of the rarest of Chinese trees. A short steep ascent, then a rather drawn-out descent, ultimately brought us to the banks of a clear-water stream of considerable size, which we crossed by an iron suspension bridge 50 yards long, and soon reached the market village of Ping-ling-shih, alt. 2900 feet. This is a small and dirty place of about 50 houses, situated on the left bank of a stream which joins the Ya Ho, some 10 li below Yachou Fu. It is in Hung-ya Hsien, in full view of Mount Wa-wu, and the most important place in the Laolin (Wilderness), as this region is denominated.

The flora of the day's journey was rather more interesting than heretofore. Wooded knolls are the rule. Evergreen trees, more especially Oak and Castanopsis, are very general, and of large size. I gathered four species belonging to the latter genus, all handsome umbrageous trees. A fine specimen of the curious Hazel-nut (*Corylus heterophylla*, var. *crista-galli*), 60 feet tall, 5 feet in girth, was one of the most interesting trees noted. The nut in this variety is hidden in a crested cup. The Chinese Fir is most abundant, being the only Conifer met with. The absence of Pine and Cypress since leaving the valley of the Ya River has been a most remarkable feature. The country generally is very broken, the sandstone bluffs bold, and clad with the usual jungle growth wherever trees are sparse.

In order to ascend Mount Wa-wu from Ping-ling-shih it was necessary for us to make a detour from our intended route. The summit was said to be 70 li distant, but, owing to the steep and difficult road, two days are required to cover this. We left behind all our spare gear and arranged what it was necessary to take into light loads. The road on clearing Ping-ling-shih ascends a rock-strewn tributary of the main stream, through a region given over to rice fields and cultivation generally. At eleven o'clock in the forenoon after traversing 30 li we reached the large temple of Tsung-tung-che, alt. 4000 feet, situated at the foot of the real ascent of Mount Wa-wu. This temple is built of wood, very old, and in poor repair. A priest and one attendant were in charge ; the rooms, though dingy and damp, were alive with fleas. But since there is no other accommodation between this place and the summit it was necessary to make the best of things. I had my bed arranged in a large hall where three huge images of Buddha looked down benignly upon me. During the morning occasional showers fell, but in the afternoon a steady downpour set in, which added to the cheerlessness of our roomy but dilapidated quarters.

Just before reaching the temple we passed through the hamlet of Tung-ch'ang Ho, where there is a very large iron foundry employing a considerable number of men. Iron ore is common in the surrounding mountains, and costs 12,000 to 13,000 cash per 10,000 catties. Every 10,000 catties of ore yields about 4000 catties of pig iron, which was said to be of good quality, and sells for 2500 to 3000 cash per picul of 100 catties. The smelting is done in furnaces heated by charcoal, which costs at the foundry 12 to 13 cash per catty. Most of the smelting is done during the winter, the summer months being given to the collecting of charcoal and iron ore. Large iron cooking-pans are also made here in considerable quantities.

Copper is also found in the same range as the iron ore, but on the opposite side. Formerly it was worked and smelted here, the name Tung-ch'ang signifying " copper-shop " or factory. From what I could learn the industry was abandoned some ten years or more ago when copper mining became a Government monopoly controlled by the officials. The people told me that they could not produce copper on paying lines under Tls. 35·00 to Tls. 36·00 per picul. The officials would only pay Tls. 28·00, consequently copper smelting was given up and replaced by that of iron. A hard, smokeless coal occurs in the neighbourhood, but is not much used. Altogether, this Tung-ch'ang Ho with its iron foundry, coal mines, and abandoned copper workings constitutes an interesting mining centre.

Around the temple are many fine trees of Castanopsis,

and the finest specimen of the interesting monotypic *Tapiscia sinensis* I have seen. This tree is fully 80 feet tall, with a girth of 12 feet. Many fine trees of the Kuei-hwa (*Osmanthus fragrans*) are planted in the temple grounds, and were in full flower, scenting the atmosphere all around. Near streams Alder (*Alnus cremastogyne*) is abundant, and on the hills the Chinese Fir is common.

It rained heavily all night, and a drizzle fell when we set out next morning at 6.30 a.m. This drizzle developed into a steady downpour as we advanced, and continued with increased violence the whole day. The road is atrocious from the very beginning. For the first 2500 feet there is a semblance of a track, some of it being made by laying pieces of split timber crosswise. The next 2500 feet is a rough scramble upwards through cane-brake and brushwood until the summit is reached. The ascent is up the north-north-east angle of the mountain, and though never really dangerous is always very difficult. We dragged ourselves upward by grasping shrubs, and it was a marvel to me how the coolies with their loads managed to overcome the ascent. The foothold was precarious, and it was often a case of one foot forward and two backward!

On reaching the summit we followed a winding path for 12 li to the temple of Kwanyin-ping, alt. 9100 feet. The mountain-top is undulating, park-like, and covered with an impenetrable jungle of Bamboo-scrub about 6 feet tall, arising from a floor of Sphagnum moss. Silver Fir (*Abies Delavayi*), called Lien sha, *i.e.* Cold Fir (signifying that it is only found in cold regions), is scattered through in quantity, but I saw no really handsome trees, all of them showing the effects of windstorms, age, and decay. The pathway across the summit is about 2½ feet wide, paved throughout with split timbers, though here and there fallen Silver Fir trees, slightly notched and flattened, have been utilized in making this roadway. We passed three temples in absolute ruins, but saw no signs of life of any description. The heavy rain and dense mists obscured all views, and I saw nothing of the country or scenery except what was encompassed in a perspective of 30 yards. Drenched to the skin but mildly describes the plight

VILLAGE OF PING-LING-SHIH, MT. WA-WU IN DISTANCE

in which we reached the temple. Our gear arrived equally wet some two hours afterwards, and we were some time getting things dry and shipshape.

The temple of Kwanyin-ping is very large, with many outhouses, and is built entirely of wood. It contains many scores of idols, but is in a poor state of repair. The main road hither is from Yungching Hsien, distant 120 li. During the Chinese fifth and sixth moons (June, July) some two to three thousand pilgrims visit this temple, but for the rest of the year it has scarcely a visitant. The priests reside at Yungching Hsien except at the pilgrim season, a novice being left in charge. This novice lives all alone, without even a dog for a companion. As a reward he receives 1½ catties of rice per diem as rations and 2000 cash (say, half a crown) per annum salary! In spite of his lonely life, and he has been in charge for three years, this novice was a very cheery person. He moved around quickly, had a ready smile, and chanted hymns and prayers wherever he went. He speedily made a fire for us to dry ourselves and clothing, and made himself generally useful. His cheery influence made itself felt, and my men soon ceased their grumbling over the vileness of the road and my madness in wanting to visit such a place. The novice told us that the first temple was built on this mountain during the Eastern Han Dynasty (A.D. 25–87). At one time there were as many as 40 temples here, but during Ming times the majority were destroyed, and the temple ornaments melted down. To-day there are only two in any sense habitable, and in one only is a man kept the year round. This same authority vouchsafed the information that the heavy rains were due to the felling of timber ; the country folk holding this view were opposed to further cutting, but the Magistrate at Yungching Hsien pooh-poohed the idea, and insisted on the slaughter being continued, with the result that torrential rains fell every day except in winter, when snow took their place.

The next morning opened dull and threatening, but eventually the sun came out and we enjoyed a fine day. The temple stands in Hungya Hsien, and is situated on the edge of a precipice. The views looking north-east over the Ya Valley and west to the Thibetan alps are very fine ; some almost

vertical limestone cliffs near by the temple are covered in a
remarkable manner with Silver Fir. The whole surroundings
are wildly romantic, and it is small wonder that the place is
deemed sacred and holy.

Wa-wu shan or Wa shan, as it is much more frequently but
erroneously called, is one of three sacred mountains, forming
the three corners of and enclosing a triangular tract of wild,
sparsely inhabited country known as the Laolin (Wilderness).
On even the most recent maps the term Lolo is written across
this region, but as a matter of fact no Lolos live here. The few
people found here are Chinese—peasants, charcoal-burners,
miners, and medicine-gatherers. The other two mountains,
Omei shan and Wa shan, have been described by former
travellers, but, with the possible exception of some Roman
Catholic priest, my visit was the first undertaken by any
foreigner to the summit of Wa-wu shan.

Like its sister mountains, Wa-wu is a gigantic upthrust of
hard limestone, but of lesser altitude than they, being only
9200 feet above sea-level. It is a huge oblong mass, composed
of a series of vertical cliffs 2000 feet and more sheer, reared on a
base of red sandstone rocks. The summit is flat with sand and
mudstone shales scattered about, and is said to be 60 li long by
40 li wide, but this is an exaggeration—30 li by 15 li being,
probably, nearer the truth. Its appearance from a distance
has already been given, and the nearer the approach the more
impressive become the perpendicular walls of rock. The
similarity in appearance between this mountain and the real
Wa shan has also been alluded to, and I strongly suspect that
the mountain seen from the summit of Omei shan and called
Wa shan is really this Wa-wu shan. Their extraordinary
vertical sides and flat summits make these two peaks unique
among the mountains of Western China.

From a botanical standpoint Mount Wa-wu proved dis-
appointing. In the first place, its altitude was some 1500 less
than I had hoped for. Secondly, all the mixed timber has been
felled for making charcoal and other purposes, leaving only a
dense shrubbery in which variety is not great. Thirdly, the
paucity of *Coniferæ* on the summit other than Silver Fir and
the impenetrable thickets of slender Bamboos which render any

extended exploration impossible. The flora generally is that common to every mountain in this region of similar altitude, but, of course, it has a certain number of species peculiarly its own in the same way as every other mountain in China has. The outstanding feature is its wealth of Bamboo-scrub ; its speciality, the abundant carpet of Sphagnum moss on the summit. This moss occurs on Wa shan and virtually on all the other mountains of this region, between 8000 and 11,500 feet, but nowhere have I seen it so luxuriantly plentiful as on Wa-wu shan.

The day being fine and clear I obtained good views of everything. The summit is made up of low, wooded hillocks, tiny dales, and glades. Here and there it is a morass, and on one occasion from such a place we flushed a Solitary Snipe. The feathery Bamboo-culms are very beautiful, and the scattered, often sentinel-like, old trees of Silver Fir quite picturesque. A few trees of Hemlock Spruce occur, but their number is infinitesimal. Some of the Silver Fir were 100 feet tall, and 10 to 12 feet in girth, but all such trees contain much dead wood. Here and there saplings are common, but they can scarcely compete with the Bamboo in the struggle for possession. At one time Davidia (both hairy and glabrous-leaved forms), Tetracentron, Magnolia, various species of Acer, Pyrus, Castanopsis, Evergreen Oak, and *Laurineæ* covered the lesser slopes, but, to-day, these are all represented only by bushes which have sprung up from the felled trees. Rhododendrons are fairly numerous, and I noted about ten species. One of these forms a tree 25 feet tall and 3 to 4 feet in girth. (It proved to be new, and has been named in honour of the Rev. Harry Openshaw, of Yachou Fu.)   Various Araliads are plentiful, and were mostly in ripe fruit. The Chinese Fir ascends to 4500 feet altitude, and very few of the evergreens other than Rhododendron extend above 6000 feet. Herbs, of course, occur, but none of any great value or interest.

A local industry of considerable importance at the season of the year my visit occurred, and for six weeks previously, is the collecting and preparing of young Bamboo shoots for culinary purposes. The species in request is one having culms the thickness of a man's thumb, and growing 10 feet tall.

The young shoots are culled when 8 to 12 inches above the ground, stripped of their sheaths and apices, leaving only the white, brittle succulent central part. These are boiled in water, then removed, suspended from rafters in a close chamber and dried by means of heat from steady-burning fires fed from locally made briquettes. When thoroughly dry they are packed in bales and carried to Chengtu and other cities, where they are esteemed a great delicacy. We saw fully a score of rude shanties where this industry was in full swing. On the spot the raw shoots are bought for 6 cash per 16-oz. catty, the collecting being done by contract. The prepared article, known as " Tsin-tzu," sells at Ping-ling-shih for 8 to 9 Tls. per 100 catties of 20 oz. each. This region is famed far and wide for its product of dried Bamboo shoots, and the industry affords employment for a large number of people.

Many wild animals, including Budorcas, Serow, Goral, Leopard, and Bear were said to occur on Wa-wu, but hunting them would be almost an impossibility. We saw no animals of any kind, but I do not doubt the reports given as to their presence on this jungle-clad mountain.

A day sufficed for our investigations, and leaving the next morning about nine o'clock, a hard day's march brought us back to Ping-ling-shih at 5.45 p.m.

Our object being to traverse this Laolin country through its greatest width to some point in the valley of the Tung River, we readjusted our loads, and the following day continued our march. Crossing the tributary stream by a rickety iron suspension bridge, we soon left Ping-ling-shih behind. The path ascends the right bank of the main stream frequently high above its waters, and at times some little distance removed. As soon as it enters limestone country the river becomes gorged. The li were long, the road rough, and it took us five hours to cover 30 li to Yüeh-ch'a-ping. This place consists of a single house, situated near where the stream bifurcates. One branch and a companion roadway leads off in a south-easterly direction, and by this track it is possible to reach Huang-mu ch'ang. The path we followed ascends the branch which swings round from the south-west, skirting the base of Wa-wu shan. After crossing a cultivated shoulder we plunged

VIEW FROM TEMPLE ON SUMMIT OF MT. WA-WU, CLIFFS CLOTHED WITH SILVER FIR
(ABIES DELAVAYI)

into a deep, narrow gorge, traversing a difficult roadway usually high up above the stream. The scenery is very fine—steep cliffs, either bare or clothed with shrubs, on every side. Journeying slowly we reached the solitary house at Chang-ho-pa, alt. 4000 feet, about 5 p.m., having covered 50 li.

During the day's march we saw a number of interesting trees, and obtained specimens and photographs. *Carrieria calycina*, a widespreading flat-topped tree, is very common in rocky places by the stream-side, and was laden with its torpedo-shaped, velvety-grey fruit which was not ripe. The Tapiscia is fairly numerous, but the trees are of no great size. Perhaps the most noteworthy tree of this region is *Meliosma Kirkii*, which has a shapely port, rigid branches, and handsome pinnate leaves, 2 feet long. Evergreen Oak, various *Laurineæ*, tall-growing Bamboos, and a Fan Palm (*Trachycarpus excelsus*) are abundant, denoting a mild, moist climate. The Chinese Fir is the only Conifer. The quantity of this useful tree and the many fine and shapely specimens were among the leading features of this trip. We had left rice behind at last, and entered a region where only maize is grown. Every available bit of land is under cultivation, but the district is very sparsely populated. A certain amount of tea is grown around Ping-ling-shih, but the industry is of little importance commercially.

The people at Chang-ho-pa informed us that the road before us was much worse than that which we had traversed. For the first 10 li after leaving our lodgings I thought they had dissembled, but afterwards the truth of their statement was only too evident   The stream flows through a narrow, wild gorge or succession of gorges ; the road is either some hundreds of feet above the stream, or down by the water's edge. The " ups and downs " repeat themselves with monotonous and irritating frequency. The path is very much overgrown with weeds and brush, always very narrow, the ascents and descents precipitous and difficult. It is misleading and foolish to term it a " road." Goats would make a better pathway, did they travel it frequently !

The scenery is grand, though mists and a drizzle of rain did their best to rob us of its enjoyment. The cliffs are in the main clothed with shrubby vegetation, but alongside the

stream large trees are common. The climate is evidently
very moist and warm, since broad-leaved evergreens abound.
Perhaps the most common shrub or small tree is a Walnut
or Chinese Butternut (*Juglans cathayensis*), which has six to
twelve fruits arranged in a raceme, and leaves up to a yard
in length. The Horse Chestnut (*Æsculus Wilsonii*), Yellow-
wood (*Cladrastis sinensis*), Hornbeam, and various Maples
are among the more interesting trees hereabouts. Clearings
and abandoned cultivated areas are overgrown with the
handsome *Anemone vitifolia*, var. *alba*, which was 4 to 5 feet
tall, and bore myriads of large attractive flowers. This herb
made a wonderful display, and I do not remember having seen
it so luxuriant elsewhere in my travels. Beneath cliffs dripping with moisture, Begonias, Impatiens, Ferns, and various
Cyrtandreæ in masses made pretty effects. The Chinese Fir
ceases at 4800 feet altitude, but limestone country is not to
its liking, and the trees quickly become scarce on quitting the
red sandstone.

Houses and patches of cultivation are few and far between,
but it is surprising that any should be found in such a precipitous country. We put up for the night at one of the three
small houses which collectively form the hamlet of Peh-sha Ho,
altitude 5000 feet, 40 li from Chang-ho-pa. The house is built
on a steep bank, overlooking a point where the stream divides,
the larger branch flowing from a southerly direction.

On leaving Peh-sha Ho we headed for the source of the
lesser of the two streams—a mere mountain torrent. Our
difficulty all day was in discerning the track and keeping to it.
I lost it early in the morning, and wasted two hours in a jungle
of Bamboo ; my Boy had the same misfortune in the afternoon.
The collecting of Bamboo shoots is an industry here as on the
other side of Wa-wu, and the tracks made by men engaged
in this are many. The path we endeavoured to follow was
frequently less well-defined than these tracks and, moreover,
was overgrown with vegetation. It crossed the torrent many
times, but the fords were difficult to discover. We passed
neither house nor person, and perforce had to explore our own
route. It rained heavily the whole day, increasing our difficulties and discomforts.

Our objective for the day was some lead mines, but early in the afternoon it became evident that we could not reach them before night was well advanced. Darkness overtook us, and we had visions of spending the night in the woods, which bound the torrent; suddenly, however, the welcome glare from a charcoal-burner's hut gladdened our hearts. Scrambling somehow down the steep slope, and across the torrent, we quickly reached this haven of shelter. It proved a wretched hovel, but the warmth from the charcoal pit was comforting since we, and all our belongings, were wet through. My bed was fixed up in a shed where prepared charcoal was stored, the men taking possession of the hut, thankful that a refuge of some sort had been found.

Much of the day's journey had consisted in struggling through brush and Bamboo, and by way of variety wading the torrent was thrown in. Whenever the mists lifted, cliffs and crags, densely covered with vegetation, were to be seen on all sides. The flora is apparently rich, but it was impossible for us to investigate it. All the larger trees have been cut down and converted into charcoal. Davidia, Tetracentron, Cercidiphyllum, and *Cornus sinensis* are common as bushy trees by the wayside; Maples are plentiful, and stout climbers, such as Actinidia, Clematoclethra, and Holbœllia are rampant.

Two men were in charge of the charcoal pits. They told us the place is called Tan-yao-tzu, and that we had only covered 30 li! All the hardwood trees having been felled they are now forced to use the softwood of Silver Fir and Hemlock Spruce, which, they said, grow in quantity on the higher crags. The charcoal is all used for smelting lead at the mines.

The roof of the shed leaked freely, but an arrangement of oil-sheets kept my bed fairly dry, and I enjoyed a good night's sleep. Awaking soon after daybreak we found it was still raining. Leaving the hut (alt. 7250 feet), we crossed two branches of the stream and scrambled up the mountain-side to rejoin the track. Soon afterwards we entered a narrow scrub-clad valley, at the head of which a precipitous, circuitous ascent brought us to the top of a ridge where the lead mines are situated. In the ascent, *Rhododendron Hanceanum* and two other species are particularly abundant, forming thickets;

Lonicera deflexicalyx is also plentiful, and was a wealth of orange-coloured fruit. On humus-clad rocks a pretty little prostrate Gaultheria with snow-white fruits is common. The hovels at the lead mines are miserable structures, but we were glad of their shelter from the rain and cold. The whole mountain appears to be full of lead, the ore (galena) being very rich. Well-shored adits are carried for considerable distances into the mountain-side, and the ore is brought out in baskets fitted on runners. The galena is pounded by hand labour into small particles ; the lead is obtained by levigation and stored in large wooden vats. Subsequently it is melted into large oblong ingots, in which form it is carried to Chengtu and Sui Fu. The freight down to the nearest waterway is very considerable. Lead has been worked in this neighbourhood for many years, and the mines are owned by a man who resides at Kiating. The labourers are paid 1800 cash per month. We were told that the previous year's output was 10,000 catties, but little reliance can be placed on this statement. Such an output is very small, but the primitive methods employed are slow and expensive. For smelting and other purposes the mountain has been denuded of its timber, and is now in its upper parts a grassy, scrub-clad wilderness. I made the altitude of the mines 9400 feet, that is to say, 2000 feet above the charcoal pits whence the fuel necessary to melt down the lead is drawn. The sides of the workings are bare and gravelly, and were covered with rich yellow flowers of a Sedum-like plant, which was new and is unknown to me.

On leaving the lead mines and crossing a slight dip we reached a babbling brook which forms the roadway for the next few li. On deserting this we made a very steep ascent to the top of a grassy ridge, alt. 10,400 feet, only to find that a deep ravine separated us from the watershed proper. After a most precipitous descent of 1600 feet over a rocky and difficult path, we reached the bed of a torrent, which I take to be the stream we noted at Peh-sha Ho flowing from a southerly direction.

On reaching this stream the rain ceased, the mists cleared away rapidly, and the sun showed itself for the first time in four days. The surrounding country is savage, and is made up of

BAMBOO JUNGLE AND SILVER FIR (ABIES DELAVAYI)

a magnificent series of limestone cliffs, their steepest crags clothed with weather-worn trees of Silver Fir. Everywhere else the trees have been cut down.

From the torrent we struggled up a severe ascent of 1000 feet, and reached the summit of the watershed, alt. 10,100 feet. Here we got a very fine view of the country, which is simply a succession of cliffs and crags capped by rugged trees of Silver Fir, and with a dense growth of broad-leaved trees in the more inaccessible pockets.

The rest of the day's journey was all downhill over a vile pathway. We reached the tiny hamlet of Yang-tientsze, alt. 7600 feet, at 6 p.m., having occupied eleven hours in covering 30 li. Two men who carried our food-stuffs arrived just as darkness closed in, and reported the rest of our gear far behind. Our lodgings were poor enough in all conscience, but most acceptable after such a fatiguing tramp. After dinner I tried to sleep on an oil-sheet spread over one of the native beds, but was soon discovered by hungry, tormenting fleas, and, tired as I was, sleep proved impossible. About one o'clock my bed and some other gear arrived. The carriers had been forced to wait after darkness fell until the moon was up in order to see the path. I could not complain; they had done their best over a most heart-breaking road. The rest of our loads turned up soon after daybreak, and we left Yang-tientsze at 7.30 a.m. Descending by a comparatively easy road for 30 li we reached before noon the village of Malie, alt. 5300 feet, a very poor place, situated on the main road between Omei Hsien and Fulin via Wa shan.

Thus had the Laolin been crossed from north-east to south-west, and, personally, I have no desire to repeat the journey. The continued rains increased considerably the difficulties of the bad roads and made what, under the most favourable weather conditions, must always be a fatiguing journey, an exceedingly arduous and miserable one. The rain and dense mists robbed the trip of its greatest charm, namely, the scenery. Except on odd occasions I saw nothing outside a radius of 50 yards. The unpropitious weather also prevented any investigation of the flora other than that alongside the pathway. In so far as it came under my observation this region possesses very little in the way of woody plants beyond what are common to the same altitude everywhere in western Szechuan. For richness in species it does not compare favourably with Mount Omei or Mount Wa. However, there are some points of interest. The region evidently enjoys a warm, wet climate, and the belt of broad-leaved evergreens, especially Oak and *Laurineæ*, extends to a greater altitude than usual. The abundance of Chinese Fir and such interesting trees as Davidia, Tetracentron, Cladrastis, Magnolia, Æsculus, Cercidiphyllum, and Chinese Butternut (*Juglans cathayensis*) is perhaps the outstanding feature. Strong-growing climbers such as Holbœllia, Actinidia, and Clematoclethra abound, and I obtained seeds of several species. Many kinds of Sorbus with white, red, and purple fruits occur, and seeds of these were also secured. Honeysuckles, Brambles, and Rhododendrons are also abundant. The scarcity of Birch, Beech, deciduous Oak, and Sweet Chestnut, and the entire absence of Pine, Cypress, and Poplar are marked features of the region. Throughout the higher altitudes Silver Fir and Hemlock Spruce are the only Conifers, although in one place I thought I detected some Spruce trees high up on the cliffs. I saw no fine trees of either of these Conifers; all that now remain grow on the crags and other equally inaccessible places, and have suffered much from the winds and weather generally. The jungle growth of Gleichenia on the sandstone, and the impenetrable Bamboo thickets everywhere between 6000 and 10,000 feet altitude, are the most striking floral characteristics of the entire region. The mining industries have been the cause of the wholesale felling of the timber.

The entire absence of decent roads, the sparse population, wretchedly poor accommodation, the savage cliffs, and jungle-clad mountain-sides sufficiently entitle this region to be termed " Laolin," *i.e.* a " Wilderness."

CHAPTER XIX

WA SHAN AND ITS FLORA

THE sister mountain to the sacred Omei is Wa shan, situated about long. 103° 14′ E., lat. 29° 21′ N., six days' journey (roughly 80 miles) from the city of Kiating. The intervening country is very rough, wild, and mountainous. The road is execrable. Baber, the first foreigner to visit and ascend this mountain, as well as Mount Omei, gives its altitude as 10,545 feet above the sea-level, 4560 feet above the neighbouring valleys. My readings were 11,250 feet above the sea, 5150 feet above the surrounding country. Allowing for error in the barometer, I think the mountain cannot be less than 11,000 feet. The flora—always a fair guide as to altitude—proves it to be higher than Mount Omei (10,800 feet) ; and this agrees with the opinion of the natives, who assert that it is the higher of the two mountains.

As seen from the top of Mount Omei it resembles a huge Noah's Ark, broadside on, perched high up amongst the clouds. Viewed from a near distance it is seen to consist of a succession of tiers of vertical limestone cliffs, only seriously broken at one point, with a peculiarly flat summit. From the hamlet of Ta-t'ien-ch'ih (6100 feet), which is situated in a depression at its base, the mountain is remarkably square looking, its four sides being more or less perpendicular. It appears to be no more than 2000 feet above the hamlet, and yet it is really 5000 feet higher. When it was first pointed out to me, 20 miles or so distant, I could not believe it was Wa shan—it looked so like a huge precipice, its massiveness belittling its height.

As already stated, the first foreigner to visit Wa shan was the late E. Colborne Baber, who made the ascent on 5th June 1878. The description of this mountain, given by him, is so accurate and beautiful that I cannot do better than quote it : " The

upper storey of this most imposing mountain is a series of twelve
or fourteen precipices, rising one above another, each not much
less than 200 feet high, and receding very slightly on all four
sides from the one next below it.   Every individual precipice
is regularly continued all round the four sides.   Or it may be
considered as a flight of thirteen steps, each 180 feet high and
30 feet broad.   Or, again, it may be described as thirteen layers
of square, or slightly oblong, limestone slabs, each 180 feet thick
and about a mile on each side, piled with careful regularity
and exact levelling upon a base 8000 feet high.   Or, perhaps, it
may be compared to a cubic crystal, stuck amid a row of
irregular gems.   Or, perhaps, it is beyond compare.   Some day
the tourist will go there and compose ' fine English ' ; he could
not choose a better place for a bad purpose ; but if he is wiser
than his kind he will look and wonder, say very little, and
pass on."

It was on the afternoon of 30th June 1903 that I arrived
at the scattered hamlet of Ta-t'ien-ch'ih, from whence the
ascent can be made.   This tiny hamlet is situated in an oval
depression, locked in by high mountains on all sides.   The
depression is about a mile long and rather less than half a mile
broad at its widest point, a small lake surrounded by a luxuriant
greensward occupies the lower end.   A species of Delphinium,
with lovely blue flowers, is very abundant.   The Chinese call it
" Wu-tzu," and say that it is poisonous to man and cattle alike.
Around the farmhouses, maize, peas, beans, buckwheat, and
Irish potato are cultivated.   The people here mostly profess
Christianity, and a Roman Catholic mission-house is the only
decent building in the hamlet.

Having procured a guide, I left the inn at 5.45 a.m. on 1st
July, to ascend the mountain.   Mists obscured everything as
we set out, and it felt very raw and cold.   The path is the merest
track—very sinuous, steep, and difficult.   Rain commenced at
2.30 p.m., and continued during the whole of the descent.
We reached our inn at 6.30 p.m., drenched through and through.

At one time a dense forest of Silver Fir covered the mountain,
but this has long since been felled, and the majority of the trees
still lie rotting where they fell.   It is a common sight to see
bushes of Rhododendrons, 20 feet or more tall, growing on the

WA SHAN, 11,200 FT.

rotting trunks. Some of these Firs could not have been less than 150 feet in height and 20 feet in girth. On the summit there are still a number of trees left, but none of great size, and nearly all have their tops broken off, either by the wind or by the snow. This mountain, in common with others I have visited, shows only too plainly the destructive nature of the Chinese. Fifty years more, under the present regime, and there will not be an acre of accessible forest left in all central, southern, and Western China. The making of charcoal alone imposes a very heavy toll on hardwood trees and shrubs. The preparing of potash salts is a common industry on the mountains west, and is another means of clearing away the vegetation in a ruthless manner. It is to the charcoal-burning industry that I attribute the marked absence of Oak, Beech, and Hornbeam.

Besides the Silver Fir (*Abies Delavayi*), the only other Conifers are *Tsuga yunnanensis*, *Juniperus formosana*, and *Picea complanata*. Rhododendrons constitute the conspicuous feature of the vegetation, and their wood is, luckily, not esteemed for making charcoal. They begin at 7500 feet, but are most abundant at 10,000 feet and upwards. In the ascent I collected 16 species. They vary from diminutive plants 4 to 6 inches high, to giants 30 feet or more tall. Their flowers, also, are of all sizes and colours, including pale yellow. It was most interesting to watch the displacement of one species by another as we ascended. One of the commonest species is *R. yanthinum*, which has flowers of various shades of purple.

The ascent of the mountain commences 100 yards or so from the inn; cultivation ceases at 6200 feet. Above this, for 1000 feet, is a belt, which has at some time been cleared for cultivation, but is now densely clad with coarse weeds. Among these occur quantities of *Rodgersia pinnata*, var. *alba*, *Spiræa Aruncus*, Astilbe, and Pedicularis, with a few bushes of *Deutzia longifolia*, *Philadelphus Wilsonii*, and Poison Ivy (*Rhus orientalis*) interspersed. Above this, for 500 feet, comes a wellnigh impenetrable thicket of Bamboo scrub. The species (*Arundinaria nitida*) is of remarkably dense growth, with thin culms, averaging 6 feet in height. Next above this, till the plateau is reached, is a belt of mixed shrubs and herbs, conspicuous amongst which are *Syringa Sargentiana*, *Hy-*

drangea anomala, H. villosa, Neillia affinis, Dipelta ventricosa, Ribes longeracemosum, var. *Davidii, Enkianthus deflexus, Styrax roseus,* Deutzia (2 spp.), Rubus (5 spp.), Viburnum (4 spp.), Spiræa (4 spp.), Acer spp., Malus spp., Sorbus spp., *Meconopsis chelidonifolia, Fragaria filipendula, Lilium giganteum,* and the herbs of the lower belt. A few Rhododendrons occur chiefly on the cliffs.

The plateau (8500 feet) is about half a mile across, marshy in places, and densely clad with shrubby vegetation and Bamboo scrub. In addition to those already noted as occurring in the belt below, we here found *Hydrangea xanthoneura, Rosa sericca,* and *Aralia chinensis,* also a species of Caltha and a few Conifers. Rhododendrons become more abundant as we advanced. Crossing this plateau we reached the north-west angle of the upper storey, and scrambled upwards by a narrow, rocky, tortuous path through dense thickets of mixed shrubs, which gradually give place to Rhododendrons as the narrow ledge at 10,000 feet is reached. *Rosa sericea,* which was past flowering below, was here a mass of lovely white. Two or three species of Lonicera and various *Labiatæ* occur within this belt, and on shady rocks at least three species of Primula, including *P. ovalifolia.*

From 10,000 feet to the summit of the mountain Rhododendron accounts for fully 99 per cent. of the ligneous vegetation. A few Conifers, Lonicera, *Rosa sericea, Clematis montana,* var. *Wilsonii,* Pieris, and Gaultheria make up the remaining one per cent. Of the herbs, Primula is the most noteworthy. Five fresh species of this genus occur, and amongst them, though uncommon, the lovely yellow-flowered *P. Prattii.* A blue-flowered Corydalis, *Cypripedium luteum,* with large yellow flowers ; *Rubus Fockeanus* and another herbaceous species are other pleasing plants. On shady rocks the curious *Berneuxia thibetica* abounds. This interesting plant was first referred to the genus Shortia by Franchet, and was later made the type of a new genus by Decaisne. The flowers are small and insignificant, white or pale pink in colour. On bare rocks I gathered the pretty white-belled *Cassiope selaginoides.*

My attention and interest, however, were chiefly taken up with the Rhododendrons. The gorgeous beauty of their flowers defies description. They were there in thousands and hundreds of thousands. Bushes of all sizes, many fully 30 feet tall and more in diameter, all clad with a wealth of blossoms that almost hid the foliage. Some flowers were crimson, some bright red, some flesh-coloured, some silvery-pink, some yellow, and others pure white. The huge rugged stems, gnarled and twisted into every conceivable shape, are draped with pendant Mosses and Lichens, prominent among the latter being *Usnea longissima*. How the Rhododendrons find roothold on these wild crags and cliffs is a marvel. Many grow on the fallen trunks of the Silver Fir and some are epiphytic. Beneath them Sphagnum moss luxuriates and makes a pretty but treacherous carpet. On bare exposed cliffs I gathered two diminutive species of Rhododendron, each only a few inches tall, one with deep purple and the other with pale yellow flowers.

Dense mists obscured our view, though about ten o'clock the sun broke through and made a temporary rift in the clouds of mist, disclosing a scene which made us hunger for more. In one place we leant over a precipice and could hear the roar of a torrent some 2000 or 3000 feet below. Near the summit three precipices, each 40 or 50 feet in height, have to be ascended by means of wooden ladders. Up these I carried my dog, never thinking of the descent. On returning he got frightened, and though we blindfolded him, he struggled hard, and on one occasion his struggles all but upset my balance. I was heartily thankful when safe ground was reached. It requires all one's nerve to mount a ladder with no balustrade, fixed to a vertical cliff 40 feet high, and on either side a yawning abyss lost in the clouds. It is at 10,700 feet—a narrow ridge not 8 feet broad—that the first ladder is encountered. From here to within a few feet of the summit the path is terribly steep, difficult, and dangerous. On clearing the topmost ladder and the remains of another, we unexpectedly reached the summit by the easiest path imaginable—for all the world like a woodland path at home.

The summit is a slightly undulating plateau, many acres in extent, with thickets of tall Rhododendrons festooned with *Clematis montana*, var. *Wilsonii*, and clumps of Silver Fir, the remnant and offspring of giants which once clothed this magnificent mountain alternating with glades carpeted with Anemones and Primulas and tiny streamlets meandering hither and thither. Baber aptly describes it as " the most charming natural park in the world."

In times past several temples existed on the summit, of which ruins only now remain. At present there is but one temple, which contains an image of P'u-hsien Pu'ssa seated on a plaster elephant. It is built of the timber of the Silver Fir (*Abies Delavayi*) and was in excellent repair. Near the temple a small patch of medicinal Rhubarb, a few cabbages, and Irish potatoes are cultivated.

The partly shrubby *Sambucus adnata* and several herbs, including Pedicularis, Microula, *Fragaria filipendula*, and *F. elatior*, range from base to summit. *Fragaria filipendula* is a new Strawberry worthy of note ; the fruit is red, more or less cylindrical in shape, often an inch in length, and of very good flavour. It is widely distributed in Western China, and at Tachienlu I have enjoyed many a dish of this fruit with cream from yak's milk.

Two days later I ascended a lofty spur (10,000 feet) of this mountain and added several new plants to my collection. Of these I may mention *Pæonia Veitchii*, *Rubus tricolor*, *Clematis Faberii*, *Ribes laurifolium*, *Potentilla Veitchii*, *Pyrola rotundifolia*, *Styrax Perkinsiæ*, *Aristolochia moupinensis*, Acer, Anemone, Pyrus, Sorbus, Berberis, and Primula. High up on the cliff *Leontopodium alpinum* and several species of Anaphalis abound. Amongst the Sphagnum at least three species of Lycopodium occur. On dripping, shady rocks and trunks of the Rhododendrons, a filmy Fern (*Hymenophyllum omeiense*) is abundant.

During the four days I botanized on this mountain I added some 220 odd species to my collection. On each of these days the work was excessively hard, and " drenched to the skin " but mildly describes our condition each evening as we reached our inn. On one occasion, through treading on some loose debris, I was only saved from being precipitated over a steep cliff by the presence of mind of a coolie who happened to be near me at the moment.

Zoölogically, Mount Wa and the surrounding wilderness is particularly interesting as being one of the places where wild cattle (*Budorcas tibetanus*) are found. I saw their foot-prints only; they were nearly as large as those of a cow. Ornithologically, it is interesting as being the home of at least five species of Pheasant, including the " Blood " and " Amherst " varieties.

I have climbed and botanized on many mountains in different parts of China, some much higher than this, but none have I found richer in cool-temperate plants, and more especially flowering shrubs. Altogether, with its rich flora, peculiar fauna, its singular geological formation, and its magnificent natural park on the summit, Wa shan has many claims on the attention of the naturalist.

END OF VOLUME I

Printed by
MORRISON & GIBB LIMITED
Edinburgh

Printed in the United States
By Bookmasters